Managing Innovation Driven Companies

Managing Innovation Driven Companies

Approaches in Practice

Edited by

Hugo Tschirky, Cornelius Herstatt, David Probert,
Hans-Georg Gemuenden, Massimo G. Colombo, Thomas
Durand, Petra C. De Weerd-Nederhof, and Tim Schweisfurth

First published 2011 by
PALGRAVE MACMILLAN

Palgrave Macmillan in the UK is an imprint of Macmillan Publishers Limited,
registered in England, company number 785998, of Houndmills, Basingstoke,
Hampshire RG21 6XS.

Palgrave Macmillan in the US is a division of St Martin's Press LLC,
175 Fifth Avenue, New York, NY 10010.

Palgrave Macmillan is the global academic imprint of the above companies
and has companies and representatives throughout the world.

Palgrave® and Macmillan® are registered trademarks in the United States,
the United Kingdom, Europe and other countries.

ISBN 978-0-230-24590-7 hardback

This book is printed on paper suitable for recycling and made from fully
managed and sustained forest sources. Logging, pulping and manufacturing
processes are expected to conform to the environmental regulations of the
country of origin.

A catalogue record for this book is available from the British Library.

A catalog record for this book is available from the Library of Congress.

10 9 8 7 6 5 4 3 2 1
20 19 18 17 16 15 14 13 12 11

Contents

Part III Innovation

Illustrations

Figures

Tables

Acknowledgements

This book is the result of the enormous efforts of many colleagues and their numerous research collaborators. To all of them we would like to express our deep gratitude for their highly appreciated personal and professional engagement behind their contributions. Our sincere thanks are directed in particular to:

- Professor Cornelius Herstatt from Hamburg University of Technology and his colleagues Antje Baumgarten, Claudia Fantapié Altobelli, Katharina Kalogerakis, Christian Lüthje, and Christopher Lettl.
- David Probert from Cambridge University and his colleagues Rick Mitchell, Robert Phaal, and Clare Farrukh.
- Professor Dilek Cetindamar from Sabanci University and her colleagues Bülent Catay and Osman Serdar Basmaci.
- Professor Massimo Colombo from Polytecnico di Milano and his collaborators Bruno Cassiman and Larissa Rabbiosi.
- Professor Thomas Durand, from l'École Centrale, Paris.
- Professor Petra de Weerd-Nederhof from University of Twente and her colleagues Annemien J.J. Pullen, Carmen Cabello-Medina, Klaasjan Visscher, and Aard J. Groen.
- Dr. Gaston Trauffler, former senior assistant at the Swiss Federal Institute of Technology.

Our gratitude also extends to Virginia Thorp and Paul Milner at Palgrave for their most cooperative, as well as vastly constructive, collaboration.

Finally we are deeply indebted to Tim Schweisfurth, assistant to Professor Herstatt. With extraordinary personal engagement, professional knowledge, and editorial skill he handled the entire process of arranging the manuscripts according to the publisher's guidelines, which included in particular the highly troublesome redrawing and reformatting of all texts, pictures and tables.

Foreword

In the spirit of EITIM

This book articulates the spirit of the European Institute for Technology and Innovation Management (EITIM). The Institute was founded in 2000 by researchers and teachers from leading European universities who shared – and still do share – a deep concern about Europe's significant underperformance regarding innovation, compared to the US and Japan. Consequently the EITIM vision "To contribute to Europe's improved innovativeness" expresses the common motivation of the founding members to focus their professional efforts meaningfully on this issue.

In the context of this overall objective, EITIM aims to enable joint cross-national initiatives which complement individual activities in the broad field of Technology and Innovation Management. Such initiatives typically include conducting joint research projects, joint supervision and promotion of dissertations, joint publications, and joint executive seminars addressing the interests of those responsible for technology and innovation on top management boards. Currently EITIM consists of the following universities: University of Cambridge (UK), École Centrale Paris (France), Hamburg University of Technology (Germany), Swiss Federal Institute of Technology (Switzerland), University of Technology Berlin (Germany), Sabanci University (Turkey), University College Dublin (Ireland), Chalmers University of Technology (Sweden), University of Twente (Netherlands), and Politecnico die Milano (Italy).

Lessons learnt from current innovation initiatives

In order to pursue the EITIM vision in practice, conclusions have been drawn from past and current initiatives aiming at reducing Europe's critical "innovation gap." Two major "lessons learnt" resulted.

On the one side, the initiatives taken by the European Union (EU) and European Commission (EC) have been analyzed. Such initiatives include the so-called EU Framework Programs. In 2000 on the occasion of the "EU Lisbon Conference" it was – perhaps over ambitiously – decided to transform Europe within ten years into the "most competitive and dynamic knowledge-based economy in the world!" To this end the sixth and seventh programs were launched with over 17 and 72 billion Euros respectively. They focused on proposed research in knowledge areas such as health, biotechnology, IT, energy, environment, transport, socioeconomic sciences, humanities, security, and space. With respect to the enormous amount being spent to promote Europe's competitiveness, the question about the effectiveness of such programs inevitably arises. Conclusive investigations do not exist. However it lies in the nature of academic research processes that it will take considerable time – 5 to 12 years – to provide knowledge which may be transferred successfully into significantly increased sales.

Moreover, regardless of available knowledge which may qualify for creating new products, noticeable sales increase due to new products – and thus increased competitiveness – come up only under the sine qua non condition, that companies master a qualified management of technology and innovation (MOT) competence. In other words, the company-internal innovation process – being the "final mile" of the long path from initiating knowledge creation to the economically useful utilization of knowledge – represents the real challenge in order to improve competitiveness. In view of the European innovation gap, it is not contradictory to conclude that this MOT competence is far from being a matter of course:

Although this relatively young discipline of company management is under rapid evolution, it still displays a shadowy existence in textbooks and education on general management. The emergence of technology and innovation management occurred in connection with the ever increasing pace of technological change. In the past, the words "innovation" and "technology" could hardly be found in minutes of top management meetings. This is a very perilous situation. If top management decisions are taken without an underpinning technology and innovation management competence, the chances are considerable that both new innovative business opportunities as well as severe technological risks are underestimated or even not recognized at all.

It is amazing indeed that this crucial role of the "final mile" is not appropriately dealt with in current competitiveness promoting initiatives such as the EU framework programs. Instead the prevailing argument expresses the opinion: "The more available useful knowledge – the higher the competitiveness of companies," a view which is quite far from the situation in the real business world.

This leads to the first lesson learnt: it is realistic to admit that exclusively knowledge research driven initiatives per se will not bring about significant short- and mid-term innovation driven improvements to Europe's unsatisfactory competitiveness in the global markets. Instead, substantial improvements can be expected only under the essential condition that European managers take hold of their own destiny and consistently worry about the innovativeness of their enterprises.

On the other side, the general focus of the management discipline MOT has been reviewed. The rapid emergence of MOT was triggered in the 1980s in the US. This was the time when the US felt seriously threatened by Japanese technology achievements. For this reason, the leading MOT document published 1987 by the National Research Council carried the title *Management of Technology – The Hidden Competitive Advantage*.

In order to take effective counter measures, MOT was created. With the intention to achieve a higher productivity of the R&D resources deployed, some first basic tools were developed, for example to identify core technologies. At that time MOT was referred to as the "missing link" between the natural sciences and engineering and general management. As a result of this initial role – outside general management – the subsequent progress of MOT research kept its focus on managing the technology driven entrepreneurial functions such as R&D and production.

This MOT role "outside general management" is not contradicted by a further research result. In order to understand the significance of "technology" and "innovation" in the context of general management, around 30 leading text books which claim to represent state-of-the-art "general management" were analyzed. The result was quite sobering, since hardly any substantial statements and recommendations on technology, innovation and their management were to be found.

This disappointment came up again from a further analysis. An earlier literature research of top management meeting minutes of successful and unsuccessful companies had been conducted. It turned out to confirm that in these documents of unsuccessful companies the words "innovation" and "technology" did not appear. These findings may presumably not have changed dramatically in the meantime.

Both these analyses mirror a perilous situation. If top management decisions are taken without fundamental technology and innovation management competence, the chances are significant that both new innovative business opportunities as well as severe technological risks are either under- or overestimated or just not recognized at all.

Out of these facts and findings, there is the second lesson learnt: it can be seen that coping with technological change has become – and will remain – a fundamental challenge not only for single entrepreneurial functions but for entire companies as a whole. This means that innovativeness and basic technology awareness have to become practically embedded qualities throughout companies at all managerial levels. In other words: they should constitute integrated values of the company culture.

These two lessons learnt clearly affected the past activities of EITIM and will continue to do so. At first, their underlying MOT concept and its content have been fundamentally revised. In the literature sometimes referred to as "New Generation MOT," top management of technology and innovation driven companies is the primary focus. This does not mean that the content of the "old MOT" has been replaced. It rather means that conceptual extensions have been developed, which include the entire company context. This concerns for example the highly sensitive issue of company culture. In fact the "New Generation MOT" can be considered to represent a first step towards an as yet missing basic textbook such as *Managing Innovation Driven Companies*. This first step is illustrated in an EITIM related book publication titled *Technology and Innovation Management on the Move – From Managing Technology to Managing Innovation-driven Enterprises.*

Then, as a major milestone in the EITIM development, a first joint book publication appeared in 2004. It carries the title *Bringing Technology and Innovation into the Boardroom.* The content attempts to overcome the "splendid isolation" of top management bodies from technology and innovation issues by explaining and illustrating the new role to be played by the upper managerial level.

Further, in the context of this book publication, EITIM is conducting yearly seminars addressed at executives from innovation-driven companies. Typical themes include the CTO function, opportunities from "open innovation" strategies, taking far-reaching technology decisions and cultural leadership measures in order to promote innovativeness throughout the entire company.

Finally, several concerted efforts have been made in the past in order to influence the EU innovation policy towards a more realistic adaption to real life situations. In brief, concrete suggestions were made on how to meaningfully spend the enormous EU resources. Realizing the crucial role of the "final mile," the explicit promotion of the MOT quality of companies as part of the EU programs would be well justified, in parallel to financing research in attractive fields such as live sciences, new materials, and so on. The question is: How much should be spent on promoting MOT competence? As a rough calculation, just 1 per cent of the allocated amounts for the EU programs would be adequate to establish 100 centers of "MOT Excellence" throughout the EU. This would be a very modest "insurance premium" in order to provide a basis for taking decisions on "doing the right thing." Incidentally, this was the approach taken by the Japanese government.

The advances at the EU level did not bear any fruits thus far. However on the national level, encouraging progress can be noticed: in Luxemburg, together with EITIM related colleagues, discussions on the crucial role of the "final mile" started about two years ago. To our surprise, by June 29 2009, a new law on the promotion of innovation had been released, which is explicitly aimed at promoting the MOT competence of companies. Truly a breakthrough.

Content

Coming back to the current book: at its core it represents a continuation of the first EITIM book mentioned above. However it emphasizes the practical side of Technology and Innovation Management. The findings presented on selected topics are research-based, however at the end of each contribution "managerial implications" cover the recommended implementation of the research results.

The content follows the same triple structure as in the previous book: Strategy, Competence, and Innovation.

In Part I "Strategy," the initial contribution "Developing Innovation Strategies: How to Start?" presents a conclusive and straightforward procedure on how to develop an innovation strategy. It demonstrates in particular two innovative MOT concepts called "Innovation Architecture" and "Strategy Morphology."

The following contribution "Organization of International Market Introduction: Can Cooperation between Central Units and Local Product Management Influence Success?" deals with the challenge of Multinational Companies (MNC) on how to best share the responsibilities of central and decentralized local product management in order to improve local market impact.

The third contribution to Part I, this "M&A and Innovation: The Role of Relatedness between Target and Acquirer," is presenting recommendations on strategic aspects – such as technological capabilities, product mix, geographic focus, and customer base – to be observed when trying to improve the effectiveness of M&As.

The final contribution to Part I "Revisiting the Firm's R&D and Technological Ecosystem" contains an innovative approach on how to systematically analye the network of relationships of a technology-based organizational unit and to draw strategically relevant conclusions.

In Part II "Competence," the first contribution "Getting Value from Technology: A Process Approach" outlines case study based solutions on how to best integrate technological competencies into business processes.

Chapter 6, "Successful New Product Development by Optimizing Development Process Effectiveness in Highly Regulated Sectors: The Case of the Spanish Medical Devices Sector" provides findings on factors of successful new product development which challenge traditionally observed managerial practices.

Chapter 7, "Roadmapping at Printco: One Company's Experience" demonstrates – again case study based – Roadmapping to be a most powerful integrating competence for strategic innovation planning.

Chapter 8, "Performance Measurements in Supply Chain Collaborations (SCC)" outlines practice-based findings on recommended competencies on how to design a performance measurement system SCC.

In Part III "Innovation," the initial contribution "Understanding Discontinuous Technology and Radical Innovation" is focused – based on "best practice cases" – on managerial implications in order to establish a system of processes which deal with developing incremental and radical innovations.

The second article "Market Research for Radical Innovations – Lessons from a Lead User Project in the Field of Medical Products" sheds light on the so-called fuzzy front end of the innovation process, in which lead users can play an eminent role. Referring to field research findings, a four-phase process on how to involve lead users in radical innovation projects is presented.

The following chapter "Relying on Experts: How to Effectively Gather Information for Innovation Projects from Market Specialists" presents recommendations for the management of external information and the identification of market experts.

The final contribution to this part "Generating Innovations through Analogies – An Empirical Investigation of Knowledge Brokers" presents the appropriate use of analogies in order to increase not only creativity effectiveness, but also to enhance project efficiency and improve communication throughout the entire product development process.

Professor Hugo Tschirky, PhD, DBA
Swiss Federal Institute of Technology (ETH)
Zurich, October 2010

Contributors

Antje Baumgarten is a Corporate Advisor at Sydbank, Germany.

Carmen Cabello-Medina is an Associate Professor in the Department of Business Administration at Universidad Pablo de Olavide, Spain.

Bruno Cassiman is a Professor in the Strategic Management Department at IESE Business School, University of Navarra, Spain.

Bülent Çatay is on the Faculty of the Department of Engineering and Natural Sciences at Sabanci University, Turkey.

Dilek Cetindamar is Director of Competitiveness Forum, Faculty of Management, Sabanci University, Turkey.

Massimo G. Colombo is Professor of the Economics of Technical Change in the School of Management at the Politecnico di Milano, Italy.

Petra C. de Weerd-Nederhof is Full Professor in the Institute for Governance Studies, SRO Innovation & Entrepreneurship at the University of Twente, The Netherlands.

Thomas Durand is Professor of Business Strategy at École Centrale Paris, France.

Claudia Fantapié Altobelli is Professor of Marketing at the Helmut Schmidt University, Hamburg, Germany.

Clare J. Farrukh is a Senior Research Associate in the Department of Engineering at the University of Cambridge, UK.

Hans-Georg Gemuenden is Professor of Technology and Innovation Management at TU Berlin, Germany.

Aard J. Groen is Full Professor in the Institute for Governance Studies, SRO Innovation & Entrepreneurship at the University of Twente, The Netherlands.

Cornelius Herstatt is a Professor in the Institute for Technology and Innovation Management at the Hamburg University of Technology, Germany.

Katharina Kalogerakis is a Research Assistant in the Institute for Technology and Innovation Management at Hamburg University of Technology, Germany.

Christopher Lettl is Full Professor in the Institute for Entrepreneurship and Innovation at Vienna University of Economics and Business, Austria.

Christian Lüthje is Full Professor in the Institute for Marketing and Innovation at Hamburg University of Technology, Germany.

Rick Mitchell is a Visiting Fellow in the Institute of Manufacturing, Cambridge, UK.

Robert Phaal is a Senior Research Associate in the Department of Engineering at the University of Cambridge, UK.

David Probert is a Reader in Technology Management at the Institute of Manufacturing, Cambridge, UK.

Annemien J. J. Pullen is a PhD Candidate in the Institute for Governance Studies, SRO Innovation & Entrepreneurship at the University of Twente, The Netherlands.

Larissa Rabbiosi is a Lecturer in the Centre for Strategy and Globalization at Copenhagen Business School, Denmark.

Tim Schweisfurth is Senior Researcher at the Institute of Technology and Innovation Management at Hamburg University, Germany.

Osman Serdar Basmaci received his MSc in Industrial Engineering at Faculty of Engineering and Natural Sciences, Sabanci University, Istambul, Turkey.

Gaston Trauffler is a Scientific Collaborator in the Centre for Enterprise Science at the Federal Institute for Technology (ETH), Switzerland.

Hugo Tschirky is Professor of Technology Management in the Centre for Enterprise Science at the Federal Institute of Technology (ETH), Switzerland.

Klaasjan Visscher is an Assistant Professor in the Institute for Governance Studies, SRO Innovation & Entrepreneurship at the University of Twente, The Netherlands.

Part I
Strategy

1

Developing Innovation Strategies: How to Start? – A Systemic Approach Using the Innovation Architecture

Hugo Tschirky and Gaston Trauffler

Introduction

Coping with the ever stunning development of technological change and mastering a competitive innovation quality have become primordial challenges for most companies. However, keeping abreast with the omnipresent reality of technological achievements and threats requires managerial competencies which differ in many respects from traditional management of the past. In the first place technology and innovation are no longer issues of specific individuals or departments but rather of concern for the entire enterprise. Whereas in the past, Research and Development (R&D) departments often were considered to be responsible for an appropriate stream flow of innovative products, today's "best-in-class" examples of innovative companies reveal an innovation consciousness which is shared by all primary entrepreneurial disciplines such as R&D, production, marketing, and finance.

Being open-minded about new technology and innovation across the entire company is the expression of a company culture which is characterized by management and employees with natural curiosity and willingness to take risks, and above all by low company barriers of internal and external communication.

The existence of communication barriers is at first a fact which can be explained to a large extent by the naturally developed differentness of the various – separate, though complementary – entrepreneurial disciplines. In a figurative sense, this differentness is expressed by differing "languages" which are "spoken" in the various disciplines: In R&D, thinking and speaking in laws of nature, engineering principles, and mathematical rules are dominating the day's work. This "language" might be called "RanDish". In contrast, the production discipline is characterized by the colloquial language "Procish" with a vocabulary consisting of specific production technologies, throughput measures, and quality notions. In marketing and sales, the situation is different again. Here people speak "Clientish," dominated by expressions such as customer benefit, sales prices, and unique selling propositions. Finally the finance world of a company is quite different in its own way. Thoughts and operations centre all around profitability, liquidity, and margins which are basic words of "Financish."

This Babylonian confusion of tongues is an everyday fact of company life and represents the main barrier for finding consensus and taking decisions in due time. In this respect, company management is in particular challenged to establish and practice innovation management processes which – on the one hand – consider the large variety of all essential factors which possibly influence major decisions. On the other hand, the same process has to provide for mutual understanding of differing professional positions, accepting changes of such positions, and finally getting to a consensus on the appropriate singular decision to be taken.

This interplay between variety and singularity reflects one way to cope with complexity: A situation can be considered to be "complex," if at a given point in time, a large number of possible next steps that can be taken is existing. Deliberately allowing for all essential decision factors to be taken into account when starting a decision-making process reflects an intentional increase of complexity. In contrast, the subsequent process to establish consensus on the best suited decision to be taken is equal to a mastered decrease of complexity.

Successfully managing complexity is a specific challenge for innovation-driven companies. The ambition to be innovative as an individual or as a company means per se to conquer new territories of customer needs and technologies which might enable meeting such needs. Therefore innovativeness relies especially on the competence to initially increase complexity during the phase of idea gathering and then to reduce it again to a reasonable number of promising innovation projects.

Typical concepts and measures to willingly increase decision complexity in innovation-driven companies include for example so-called "business intelligence systems." They have become an agenda point dealt with priority. They aim at providing the appropriate inflow of information based on the experience that successful innovativeness relies to a large extent on knowledge which has been gathered outside the company. Such systems vary considerably in terms of dedicated resources depending on the size and the knowledge focus of the company. However there are even middle-sized and small companies which have introduced processes which systematically provide relevant outside knowledge based on a coordinated and knowledge field focused information search by single individuals, often called "gatekeepers."

A further challenge of emerging significance is the broad-minded awareness of innovation. A still popular understanding of innovation is focused on new products, new technologies or new services. However, innovative excellence goes far beyond and also includes organizational and business innovations as well. Particularly the latter ones require managerial creativity with perspectives across industries and global economies. Whereas incremental product innovations are gradually dealt with using dedicated processes, service innovations are still far from being systematically explored. And above all the deliberate handling of disruptive technologies and radical innovations is still a very rare managerial competence. However, good examples of companies which deal with this issue successfully exist.

Yet another leading trend has to be mentioned. Under the terms "open innovation" and "distributed innovation" reference is made to structural patterns that are characterized by pooling R&D resources across and beyond organizational and regional boundaries. In this respect Procter & Gamble (P&G) is providing an illustrative example. On P&G's website the chairman of the board, A. G. Lafley, is personally promoting the initiative "connect & develop": "I want us to be the absolute best at spotting, developing and leveraging relationships with best-in-class partners in every part of our business." One of the measures consists of two entries on the website, one reserved for "ready-to-go products or technology" to be offered to P&G, and the other installed for the acquisition of technologies from P&G.

Moreover the venture capital scene is providing an attractive potential for a managed increase of complexity and thus an accelerated innovative growth. The careful and systematic monitoring of start-up companies for example enables identifying the "right" technologies to be made available in a short time and with reduced technological development risks. Finally, further practices to initiate innovation search include highly interdisciplinary decision groups, external networks of experts, lead supplier, and lead user concepts of product development, etc.

Despite these ambitious challenges many professional authors have come to the conclusion that "innovation is not black art." This means that innovativeness is not the result of coincidental circumstances but rather the consequence of responsible leadership which makes use of already available tools and methods of modern management of technology and innovation.

In the following, four selected instruments of innovation management are presented: The newly developed "Innovation Architecture" (IA), the Strategy Morphology (SM), the so-called Business Roadmap (BRM), and the Business Model (BM). They are suited to effectively support the mentioned interplay between increasing and decreasing management complexity. And, above all, they represent practical management instruments aimed at enabling mutual understanding across the various company disciplines, functions, and management levels.

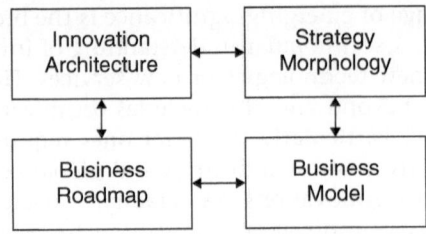

Figure 1.1   "Quadrangle" of Innovation Architecture, Strategy Morphology, Business Roadmap, and Business Model representing the core of strategic Technology and Innovation Planning

These four instruments are closely interrelated and represent the core of modern strategic Technology and Innovation Management (Figure 1.1).

Innovation Architecture representing structured creativity

Generic concept of Innovation Architecture

The IA represents a systemic knowledge map. It covers on the one side all major knowledge which underlies existing and new businesses and products in terms of trends of customer needs and available scientific knowledge (Figure 1.2). On the other side, it displays plausible combinations of such knowledge which constitute the basis of existing and new businesses and products. In this respect it consists of six structural layers: (new) Innovation Trends, (new) businesses, (new) Products-Systems-Services, (new) Product Functions, (new) Technology Platforms, and (new) Scientific Knowledge Fields.

The level of "(new) Innovation Trends" contains all relevant societal, economic, and market trends such as aging society, globalization, knowledge society, technology markets, venture capital, etc. In addition to consider such singular trends it is highly helpful also to include certified assumptions on future processes of activities of individuals, within homes and within organizations, and in society. On the level "(new) Businesses" ideas of future businesses are itemized which complement or replace current businesses. The level "(new) Products-Systems-Services" comprises the existing and possibly future products which might be part of the current or new businesses.

On level "(new) Product Functions" we find items or product features which are shared by more than one product. The implied "thinking in product functions" is absolutely key to IA: Product functions describe primary effects of products and services, such as "print characters" and "process transparencies" in the case of a printer. In this view product functions constitute "translations" of customer needs without preempting the technological

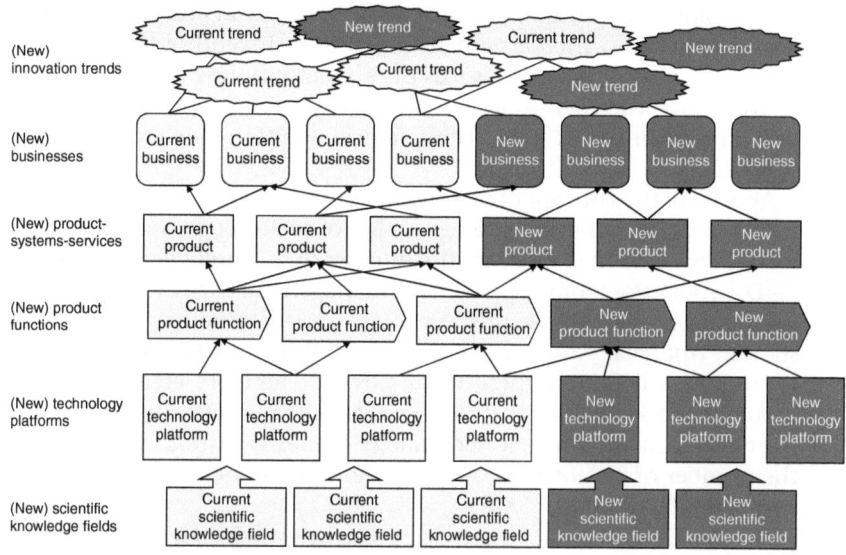

Figure 1.2 Generic concept of Innovation Architecture

solutions. This is essential in order not "to jump to conclusions" too early before all possible technological solutions for each individual functions has been investigated.

Finding best-suited technological solutions is subsequent to identifying major current and new product functions and is assigned to specific technology platforms. They are visualized on the next level of the IA. Together with product functions, "Technology Platforms" constitute the other main focus of IA: The term "Technology Platform" is a strategic term. It itemizes a group of product and process technologies which are to be mastered in order to provide competitive solutions for the identified product functions. Example: A renowned producer of diagnostics instruments identified six product functions which are shared by more than one product or system of two independent business units: "to handle objects," "to control temperature," "to process samples," "to perform analytical measuring," "to control systems," and "to communicate data." Correspondingly, separate technology platforms have been established which are focused on developing best-suited solutions of the various product functions relevant for quite different types of products and systems.

Prior to introducing the concept of technology platforms, two separate R&D departments existed which were assigned to develop pharmaceutical instruments for various business units. In this organization, each R&D department was challenged separately to develop entire "product solutions."

This meant that technological solutions for identical product functions were developed twice which is verbally equal to "reinventing the wheel" several times. Avoiding this obvious loss of productivity was the main motivation to introduce the mentioned technology platforms focused – across business unit boundaries – on technological solutions for individual product functions.

Above all, the concept of technology platforms has a strategic significance as it allows stressing the strategic value of technology: In innovation- and technology-driven companies the number of technologies to be mastered in R&D, production, and supply chain management easily amounts to several hundreds or even thousands. Although each technology represents a substantial financial value, it is practically not feasible to establish a business value for each one of these technologies. Therefore creating meaningful groups of technologies in terms of technology platforms allows management to estimate strategic technology values (Figure 1.3). In this respect the number of platforms is certainly more than one, however, ought not to exceed half a dozen for SME's.

On the final IA-level "(new) Scientific Knowledge Fields" are positioned. They indicate realistically and specifically the origin of the company's technology base. The careful treatment of this level has a significant strategic meaning as well: It reflects the "Strategic Competence Planning" as an indispensable issue of modern strategic planning. Ambitious marketing plans based on technology-intensive products remains superficial until the corresponding fundamental knowledge base has been planned and established. Example: A successful producer of specialty chemicals came to the conclusion that future innovative growth – hopefully as a result of radical innovations – must be based on knowledge from life sciences, nanotechnology, catalysis, and polymers. Having identified these fields as strategic competences to be mastered in future, five separate midterm projects were

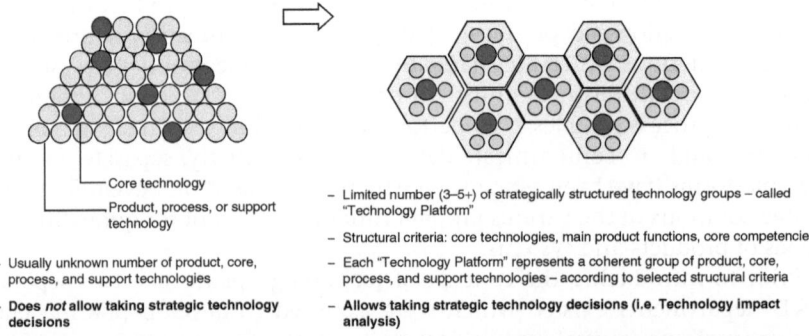

- Core technology
- Product, process, or support technology
- Usually unknown number of product, core, process, and support technologies
- **Does *not* allow taking strategic technology decisions**

- Limited number (3–5+) of strategically structured technology groups – called "Technology Platform"
- Structural criteria: core technologies, main product functions, core competencies
- Each "Technology Platform" represents a coherent group of product, core, process, and support technologies – according to selected structural criteria
- **Allows taking strategic technology decisions (i.e. Technology impact analysis)**

Figure 1.3   From an unstructured to a structured collectivity of technologies

launched in order to establish the aimed at knowledge bases. Only afterwards concrete product innovation planning and development was started. And indeed, several radical innovations using the newly acquired competences resulted.

Generating innovation architectures

Developing an IA does not have to follow strict rules. However, following an often practiced procedure, at first, the current business situation is displayed. To this end in step 1 the existing businesses are captured (Figure 1.4). In step 2 those trends are identified which are of influence for the current businesses. Step 3 deals with the main current products and their affiliation with the different businesses. Step 4 consists of bringing in the product functions which are shared by the current products and services and connecting them with these products and services.

In step 5 the current technology platforms are displayed and connected with the product functions, which they are primarily focused on. The presentation of the current business situation closes with bringing in the essential scientific knowledge fields which had a major influence on product development in the past (step 6).

The creative part as such of using the IA starts with steps 7 and 8: Resulting from earlier studies or dedicated workshops, new significant trends which presumably will influence customer behavior in future and generate new

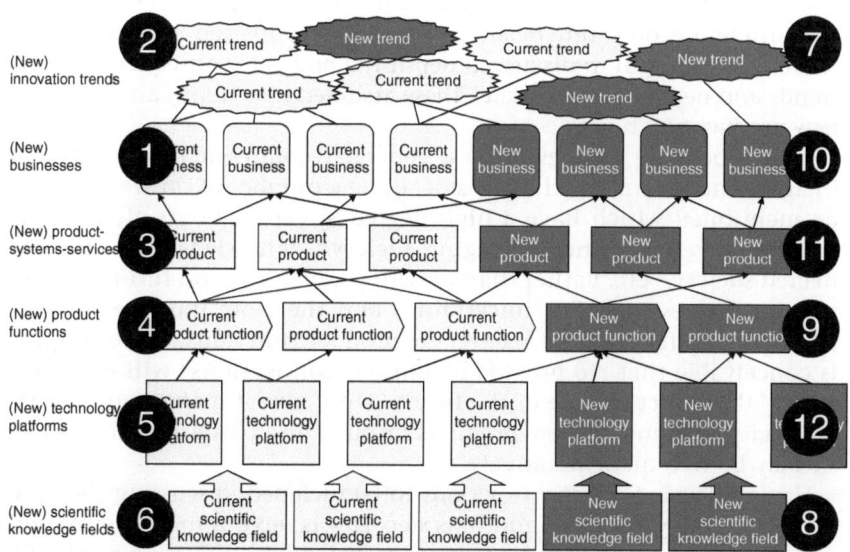

Figure 1.4 Generating an Innovation Architecture

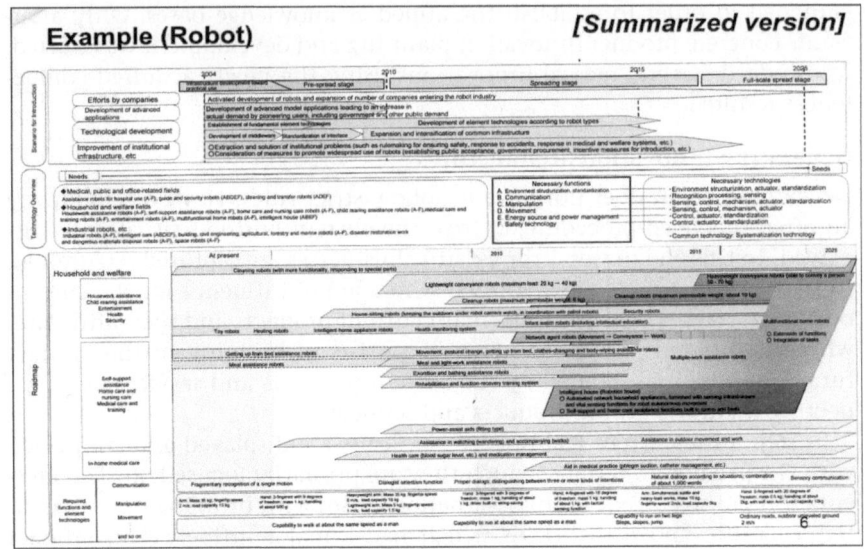

Figure 1.5 Examples of the METI Roadmap Compilation: Robotics

customer needs are introduced. Correspondingly new knowledge fields have to be brought in, in order to stimulate the minds also from this side.

As a next step, it is recommended not to follow the given IA structure and to storm brains on future businesses or products. Instead, making step 9, it is more meaningful to discuss in depth the probable consequences of new trends and needs and "translate" these assumed new trends and needs into new product functions.

This procedure may even be a general recommendation on how to imagine the future: On the one side, quite a number of future trends can be mentioned which have a high degree of certitude. As already mentioned above, they include "aging society," "knowledge society," "connected society," etc. Rather than drawing conclusions on future products and businesses, it is less uncertain – and therefore more reliable – to illustrate the future with likely new product functions. For example, it is conceivable that "to provide ubiquitous connections" will become an essential product feature of the future. Only now it makes sense to have creativity sessions on new products or businesses, whereas steps 10 and 11 may be well made iteratively.

Having decided on new functions to be included, discussing the necessary technology backup – and thus step 12 – is next. A most useful source of information for this step has been provided by METI: As a result of an extensive collaboration with industry and academia METI has elaborated a wealth of industry roadmaps of major technologies and natural sciences. They cover current and future developments and include the fields of IT, life science, environment, energy, robotic, aircraft, aerospace, nanotech, and MEMS technologies. A typical example is given in Figure 1.5.

The outcome of this step might suggest a new technology platform or the combined development of a new function together with an existing platform. Also during step 12, the question of new knowledge fields is discussed anew.

Practical example

To illustrate the use of IA in practice, the following example is given (Figure 1.6). It is a neutralized case of a company which has been active in the past in the fields of office machines, systems solutions, optical products, and cameras. Company management has decided to extend business activities into new markets. In order to carefully select new business options, some 300 engineers, physicists, and representatives from other sciences are doing detailed studies on the "world of tomorrow."

At first the identified trends have to be mentioned, which are in Japan generally considered to be of future societal influence. For example the government has published a "White Book" already some years ago, that expresses significant developments, among which the trend towards a knowledge-based society is a main theme. Also it has been decided to acquire new basic

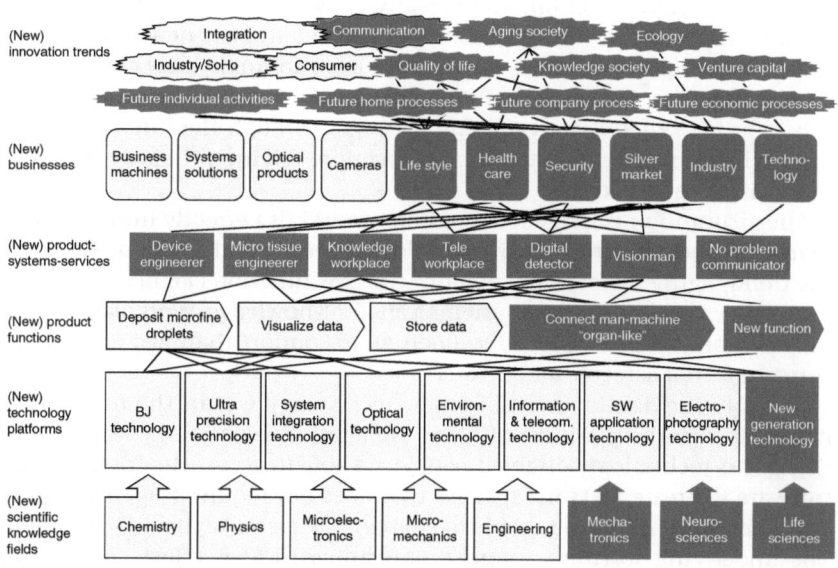

Figure 1.6   Innovation Architecture: Example from a workshop in practice

knowledge from scientific fields such as mechatronics, neurosciences, and life sciences.

Among the new functions, "connect man-machine 'organ-like'" has to be emphasized. This function "translates" the ultimate customer need to really have to deal with any kind of technology without any handling difficulties. Preceding the identification of this function, workshop participants had to answer the question: "What is more complicated: eating a sandwich or using a video recorder?" The answer to this rhetorical question was obvious, since dealing with a video recorder usually is a rather tedious affair.

Having got consensus on the fact that "eating a sandwich" is much less complex than handling the video recorder, a video was shown on the very complex process of digesting a sandwich. Therefore the next question came up: Why is the as such extremely complex process of eating a sandwich conceived by individuals to be easy and handling the rather banal technology of video recorders thought to be complex or at least complicated? The answer was obvious: Whereas the recorder technology is mostly poorly adapted to human needs, "digestion technology" is – on the contrary – perfectly adapted to human needs.

The workshop concluded by establishing a final objective for developing technology: "Technology has to fit human needs like an organ!" Based on that thesis, the above mentioned function was formulated. And, without a discussion on how to realize the new function technologically, a corresponding platform "New Generation Technology" was identified.

Using the new function, the "creativity game" within the innovation architecture was continued by combing current and the new product function to come up with all sorts of new products and finally new businesses: As an example, the product "No Problem Communicator" can be mentioned. It is a device which is transmitting messages through the internet without any active computer interaction and therefore relying primarily on speech recognition technology.

The assumption is that such a product would fit perfectly into the "Silver Market" which is directed towards elderly people and thus correspond to the trend "aging society." But equally, the new function can play the central role when developing new systems called "Knowledge Workplace" which indispensably will have to offer much more comfort than just a number of computers, flat screens, and printers.

Recently the development of a new technology was in the news, which might fit into the platform of "New Generation Technology"(Figure 1.7).[1] Using this technology, disabled people are able to access the Internet using software to create texts in Japanese that does not require fingers to tap the keyboard. All they have to do is look at an on-screen keyboard – and blink. The underlying software – named "Mitsumeru Dake" ("Just Look") – has been developed by Prof. Kohei Arai from Saga University.[2]

One other new product idea may be mentioned: the "Micro Tissue Engineerer." This product is supposed to be capable of producing living

Figure 1.7 Prof. Kohei Arai from Saga University is developing software that helps disabled to use computers by simply blinking at an on-screen keyboard

micro tissues such as skin, cartilage, and cornea, and application to other important organs is anticipated. The production method is called "bioprinting" which is researched for example by Prof. Makoto Nakamura at Kanagawa Academy of Science and Technology.[3] It is based on a bio-specialized printing procedure using a bioprinting machine which compares to color ink jet and multi layer printing: living cells and proteins are directly and precisely positioned and arranged at the respective intended position.

The principal product function for this "Micro Tissue Engineerer" has been named "Deposit Micro-Fine Droplets." In first versions of the innovation architecture the spelling of this function was "Print Characters" in order to refer to ink jet printing machines, which characterize the product range of the analyzed company. However having realized the possibility of bioprinting, the new spelling has been chosen which includes printing characters as well.

Highlights of innovation architectures

Market Pull and Technology Push

In literature on technology and innovation management, these two terms are often used. They refer to two quite different patterns of technology to become applied in marketable products (Figure 1.8).

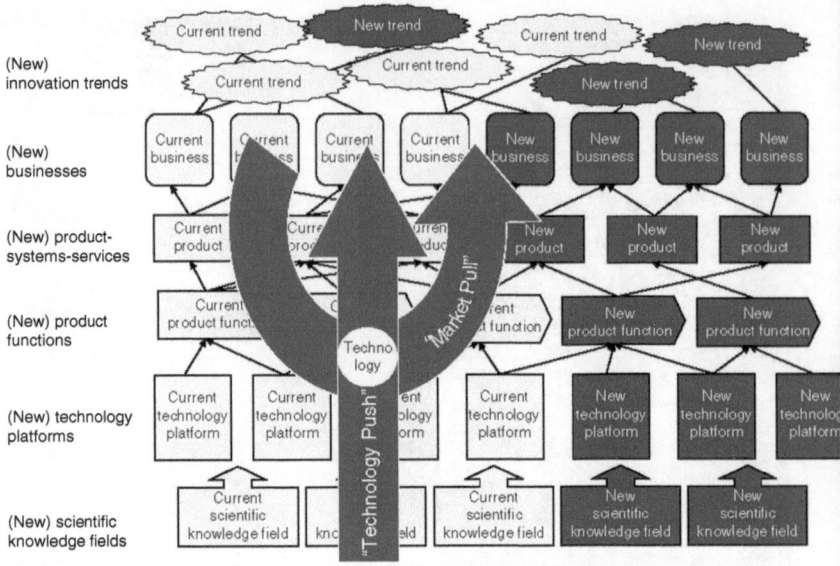

Figure 1.8 Innovation Architecture considers "Market Pull" and "Technology Push"

"Market Pull" on the one side describes situations, in which a specific market demand or customer need requires a specific technological solution in order to be satisfied. For example, the development of the 300 mm waver technology has been a response to strong trends towards production cost reduction.

"Technology Push" on the other side refers to planned or haphazardly discovered technologies, which require useful applications in the market place. "Bubble Jet" technology is such an example. The effect of ejected water bubbles jetting out of a pipette tip under influence of heat was observed in 1977 by Mr. Endo at the Canon research laboratory. His highly visionary mind made him imagine a new printing technology. After more than ten years of tireless R&D work, this vision became a reality. Today "Bubble jet" or "Ink Jet" technology is a most successful technology used in printers of renowned companies such as Canon and Hewlett Packard.

The IA allows taking into account both patterns. Be it the challenge to find product solutions for new customer needs or searching for meaningful new technology applications, both streams of analysis can be handled using the IA methodology.

Structured creativity

The IA has numerous independent entries (Figure 1.9).

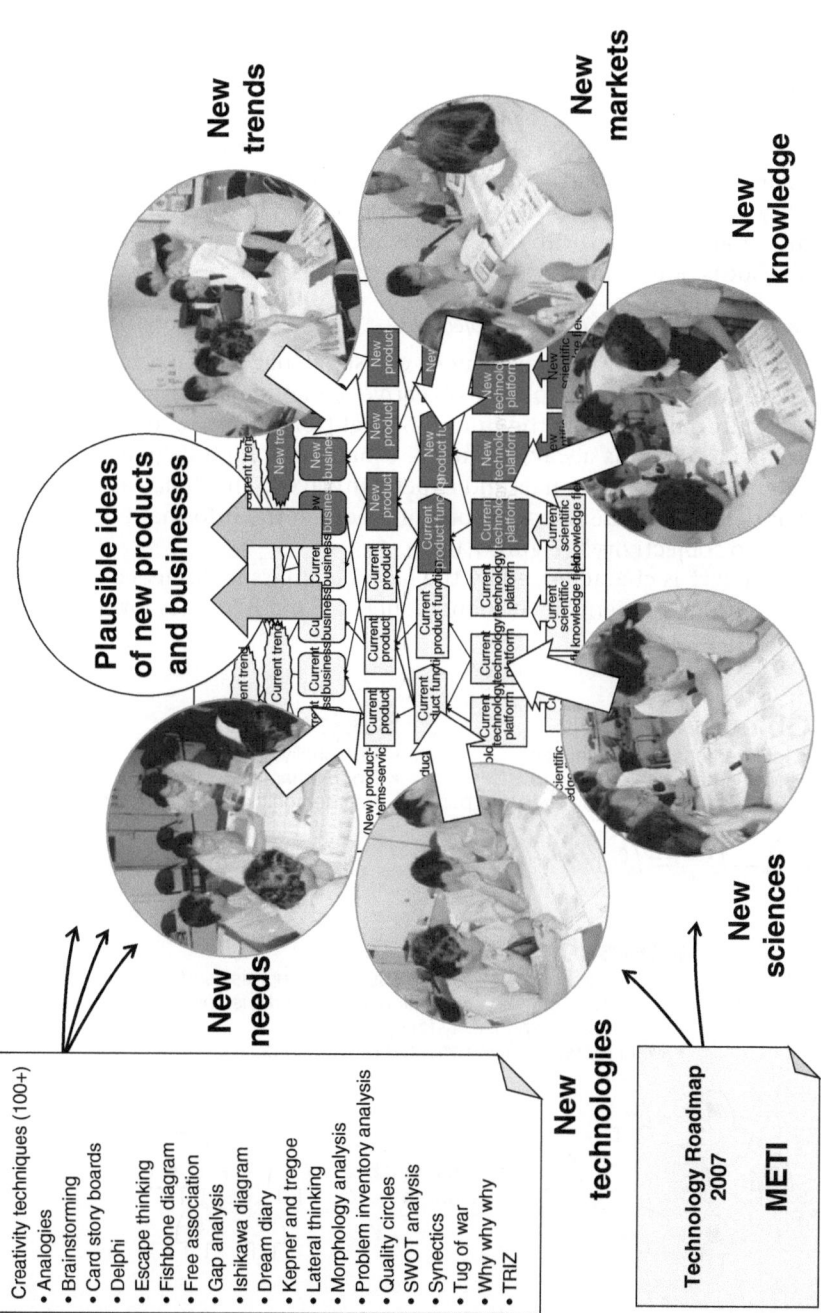

Creativity techniques (100+)
- Analogies
- Brainstorming
- Card story boards
- Delphi
- Escape thinking
- Fishbone diagram
- Free association
- Gap analysis
- Ishikawa diagram
- Dream diary
- Kepner and tregoe
- Lateral thinking
- Morphology analysis
- Problem inventory analysis
- Quality circles
- SWOT analysis
- Synectics
- Tug of war
- Why why why
- TRIZ

New trends

New markets

New knowledge

New needs

New sciences

New technologies

Technology Roadmap 2007

METI

Plausible ideas of new products and businesses

Figure 1.9 Innovation Architecture – Structured Creativity

Essential input data such as new customer needs, new societal trends, new technologies (METI Technology Roadmap (TRM) 2006), etc. typically result from separate studies or group works. For most of these, input data of one or more different creativity techniques can have been used. As a matter of fact, the more independent origin of these input data, the more valuable outcome can be expected developing the IA subsequently. This characteristic explains the notion "structured creativity": on the one hand, unlimited creativity can be applied for developing input data. On the other hand, the IA concept allows structuring these ideas meaningfully in order to obtain plausible ideas of new products and businesses.

Consensus building based on intersubjectivity

It is fundamental to any management decision, that they never ever can be based on exclusively – purely objective – mathematical calculations. Therefore, in management the question "right or wrong" is out of place. Management decisions always include – usually to a high degree – subjective judgments of professionals and experts involved. This holds also for all IA input data. Therefore the question comes up on the information quality next best to "objectivity" (Figure 1.10).

"Objectivity" is characterized by the fact that an issue discussed is based on unquestioned scientific evidence such as mathematics or physics.

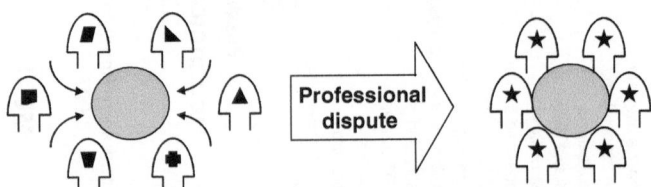

Objectivity
- Interpersonally revealed fact
- "Mathematically" evidenced
- To be demonstrated by even one person

Subjectivity
- Personally appraised fact
- No "mathematical" evidence
- Interpersonally different appraisal

Intersubjectivity
Consensus between qualified subjective appraisals of:
- Carriers of original information for taking decisions
- Carriers of responsbility for decisions
- Main representatives affected by decisions

Professional dispute

Figure 1.10   Concept of intersubjectivity

Therefore, any expert in the field comes up with the same evaluation, provided that he masters the relevant scientific knowledge. However, for reasons given above, "objectivity" is no acceptable concept for management practice.

"Subjectivity" represents the opposed information quality: due to absence of scientific tools, situational evaluations by individuals dominate. Therefore evaluations of the same situation by different professionals may very well result in as many evaluations as people involved. For that reason, "subjectivity" is certainly not a suitable basis for taking management decisions either.

For these reasons, "intersubjectivity" has become a primary canon for management decisions. It means a consensus between qualified subjective evaluations of situations or actualities. "Intersubjectivity" can be reached as a result from professional disputes. In this respect the selection of the people which are involved in such disputes is all-dominant. From a social science perspective it seems to be appropriate if at least three kinds of participants are selected: (1) professional carriers of original information for taking decisions, (2) carriers of original responsibility for taking decisions, and (3) main representatives who are affected by the decisions to be taken.

These reflections underline the strongly recommended development of IA based on multiple group work and – simultaneously – may support the constitution of the corresponding groups.

Innovation Strategy Morphology

Generic concept of Strategy Morphology

The "Morphology Matrix" represents a universal schematic for developing problem solutions. It shall be explained by solving the "problem" of making a trip to the Fuji san (Figure 1.11):

The matrix has two dimensions. The dimension "Solution Elements" consists of all essential parameters which are part of the problem solution when travelling to the Fuji san, such the time of year, way of travelling (travel transport), grouping (participants), guide, duration, and focus. The dimension "Specifications" contains for each single one of the solutions elements any conceivable expression. For example: "the time of year" may have the specifications "January-March," "April-June" etc. ; "way of travelling" may have the specifications "bus," "train" or "private car."

After having enumerated all essential solution parameters and after having enumerated all conceivable specifications of each parameter, solution options can be developed. The procedure is the following: For each one of the parameters, one specification is chosen in such a way, that all chosen specification together constitute a reasonable solution option. For example:

Solution element	Specifications			
Time of the year	○ January–March	○ April–June	○ July–September	○ October–December
Travel transport		○ Train	○ Private car	○
Grouping	○ Alone	Family	○ Travel group	○
Guide	○ No guide	With guide	○	○
Duration of trip	2 days	○ 3 days	○ One week	○
Focus of the trip	Fuji san only	○ Fuji san + Hakone	○ Fuji san + Jap. Alps	○
Solution options	**○ Option 1**	**○ Option 2**	**○ Option 3**	**○ Option 4**

Figure 1.11 Example "Morphology Matrix": trip to Fuji san

making the trip during April-June, taking a bus, going with family, taking a guide, planning for two days, and going to the Fuji san alone seems to be a reasonable option (option 1 in Figure 1.11). Correspondingly, other options can be developed.

Developing solution options has to follow two indispensable rules:

Rule 1: The chosen option has to include all parameters without exception.
Rule 2: For each parameter only one specification can be chosen.

This means: When developing specifications for the various parameters, they have to be formulated in such a way, that they are mutually exclusive.

Developing an Innovation Strategy Morphology (ISM)

The "Morphology Matrix" shall now be applied for developing options of business (innovation) strategies:

To this end at first the term "strategy" has to be defined. "Strategy" shall be understood as a "set of mid-term decisions which express the selected major business activities and emphasize the priorities of the resources to de deployed." In this view it is recommended to distinguish between "strategic objectives" and "strategic paths" (Figure 1.12).

Strategic objectives are for example: financial performance (return on equity (ROE), return on investment (ROI), etc.), external growth or internal growth, innovation rate, etc. Strategic objectives are usually inputs

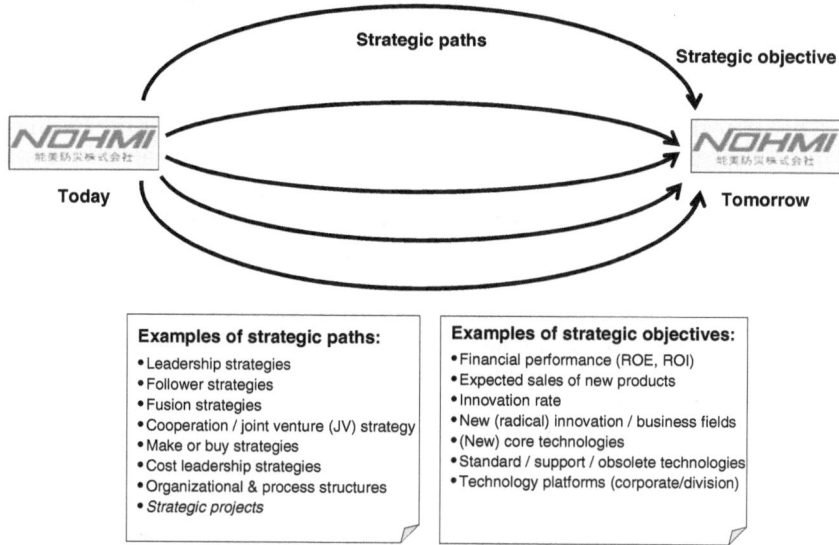

Figure 1.12   Distinction between strategic objectives and strategic paths

	Strategy elements	Specifications			
Strategy goals	Strategic purpose	○ Internal growth	○ External growth	○ Int. & ext. Growth	○
	Return in equity (roe)	○ 8 %	○ 10%	○ 12%	○
	Return on investment (roi)	○ 10%	○ 12%	○ 14%	○
	Annual sales in three years (million Ye	○ +5000	○ +7000	○ +10000	○
	Long-Term sales growth	○ 5%	○ 10%	○ 15%	○
	Innovation rate	○ 10%	○ 15%	○ 20%	○
		○	○	○	○
Strategy path	Innovation field: polymers	○	○ Polymer belts	○ Polymer cables	○ Medical polymers
	Innovation field: optical fibres	○	○ Optical switches	○ Optical data transfe	○
	Innovation field: smart materials	○	○ Acoustic attenuation	○ Smart belts	○
	Innovation field:	○	○	○	○
	Business field: polymers	○	○ Building market	○ Electronics market	○ Health care market
	Business field: optical fibres	○	○ Electronics market	○ Automotive market	○
	Business field: smart materials	○	○ Automotive market	○ Entertainment mark	○
	Business field:	○	○	○	○
	Overall market strategy	○ Leader strategy	○ Follower strategy	○	○
	Regional market strategy	○ Japan	○ Japan US	○ Japan- US+ Europe	○
	Market collaboration strategy	○ No collaboration	○ Majority partnership	○ Minority partnership	○
	Market collaboration partners	○ No partner	○ Freudenberg	○ Schlumberg	○ Company X
	Technology strategy	○ No collaboration	○ Open innovation	○ JV competence	○ New strat. Competence
	Make or buy strategy	○ In-House R&D	○ Contract research	○ Joint R&D	○
	Technology collaboration partners	○ No partner	○ University X	○ Company Y	○ Company Z
	Acquisition candidates	○ No candidate	○ Saint-Gobain	○ Company A	○ Company B
	Strategy options	○ **Option 1**	○ **Option 2**	○ **Option 3**	○ **Option 4**

Figure 1.13   Applied Strategy Morphology with resulting two strategy options

from the next higher level of management. For example: the overall strategic plan can decide to aim at ROE values of 15 percent. In that case this ROE value is a strategic objective for the business units being parts of the entire company.

The strategic paths in that case represent strategic options for the business units. They can consist of choosing between different types of new businesses, make-or-buy strategies, scope of cooperation, priorities on markets, etc.

A typical practical SM is illustrated in Figure 1.13. It shows that among the parameters of the strategic path the planning business unit had the option to decide between three innovation fields (polymers, optical fibers, smart materials) and corresponding business fields, four types of market strategy (leader or follower strategy, regional options, minority or majority partnerships, and different collaboration partners), etc.

According to Figure 1.13, it had finally been decided to focus on two strategies to be studied more in detail, namely a purely external growth strategy (option 1) and a mixed internal and external growth strategy (option 2).

Business Roadmaps, a strategic timeline for innovative business planning

Generic concept of Business Roadmaps (BRM)

The philosophy reflected by the roadmap used in strategic management is inspired from a geographic roadmap. Geographic roadmaps serve as a traveler's tool providing essential understanding, proximity, direction, and some degree of certainty in travel planning when moving from a current to a new location. In analogy, the roadmap in strategic management supports managers as a planning tool when navigating the company (or parts of it) from a present state to a future state.

As a planning tool the roadmap is in its very essence a graphical chart that displays the objects to be planned, the relationship between these objects and the dependence of the latter over time. The activities corresponding to its elaboration and updating are called roadmapping.

Roadmaps and roadmapping can take a variety of specific forms, depending on the type (opportunities, capabilities, products, technologies, etc.), the particular context (industry, science, enterprise), and the time horizon (short-term, midterm, long-term). This paper focuses on the Technology Roadmaps, which display for a given enterprise the following three main objects levels over a mid- to long-term time horizon: Innovation drivers, Businesses, and Technologies (refer to Figure 1.14).

The level of Innovation drivers includes company-external and-internal drivers influencing the planning activities within the roadmap as an initial position. External innovation drivers are factors derived from the company's external environment; they can be formulated as market or customer

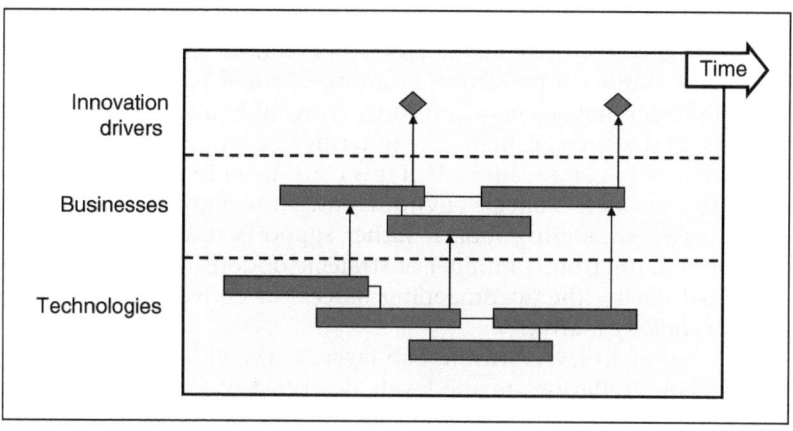

Figure 1.14   Generic concept of the strategic Business Roadmap

requirements or needs. Those requirements and needs are initiated for example by changing legal regulations, by technology change, by competition,
by social trends, etc. Internal drivers often originate from the company's
strategic goals, or from company internal constrains such as products release
politics, company strength or weaknesses, cost reduction goals, profit goals,
restructuring programs, etc.

The level of Businesses comprises the company's strategic businesses and
in its sub-layers corresponding products and services to be delivered in
those businesses. Besides products and services this level can also include
as in additional sub-layers characteristics of the actual planned products or
service.

The third level is dedicated to Technologies. This level is subdivided into a
Technology Platform layer and an actual Technology sub-layers.

All the levels and their corresponding sublevels of the Roadmap, such as
the actual drivers, technologies, products, product characteristics as well
as the content of those level and layers themselves are not to be considered
isolated but only make sense when their alignment represents a holistic and
consistent strategic plan. This alignment is given through scheduling and
interrelating all three levels and their corresponding content in a way to
meet strategic goals within set timelines.

This being said, Roadmaps do not replace strategic analysis and strategic
goal setting. Before starting with Roadmapping, strategic goals must be clear,
else the Roadmap risks to sketch a plan that has nothing to do with strategic
management. As a matter of fact, strategic goals need to be included in the
Roadmap as company internal drivers.

Furthermore, the distinction between a project and a Roadmap is important. Projects are highly specified activities, well defined in time, usually

over a shorter time period, and with a low level of uncertainty. Also the content of a project may make sense when considered isolated from other projects. Planning of projects is of an operational nature and follows the rules of project management and controlling. BRM are however used as one progress up the strategic hierarchy in terms of complexity and uncertainty. Its content only makes sense when it is considered in alignment with itself and with the whole strategic environment. Thus, a Roadmap is much more than a mere scheduling tool; it rather supports strategic evaluation and decision-making from a number of strategic options. These options can be composed during the Roadmapping process or derived from the IA using the Morphology Matrix.

The three main levels with its sub-layers of the BRM, as described above, show major similarities to the levels described in the IA. As a matter of fact, the Roadmap can be seen as one dynamic form of the IA. It is however used for a different end. The IA mainly focuses on a snapshot of a present and future state in order to enhance its author's creativity when looking for innovations. It creates transparency of a company's innovation assets and shows the relationships between. The Roadmap aims at enhancing its authors planning and decision-making capabilities by showing the dynamic relationship of its objects over time. This dynamic way of displaying the objects visualizes the strategic options of how planned goals can be achieved and what needs to be decided about.

The following section will display a method of developing the BRM and how the IA contributes to the latter.

Developing Business Roadmaps

This paper suggests a practice-proofed procedure to develop a Roadmap. It is a method to develop the BRM in practice using the IA as a basis. The IA taken as an example is the one of a company aiming to realize affordable flying vehicles for individual mobility (Figure 1.15).

Taking the above IA as an initial position, the present section will describe a step-by-step procedure to generate a BRM (Figure 1.16).

Step 0: Decide about the time horizon of the TRM. A meaningful planning horizon for a strategic TRM is mid-, to long-term rather than short-term. Most often, this represents the time range necessary to implement a technology and innovation strategy effectively. Depending on the industry, the number of years of this planning horizon varies considerably. Thus, we recommend adapting it individually according to the pace of each industry.

Step 1: Customer segmentation and Innovation Drivers: Define the customer segment for which you wish to sketch the Roadmap. A clear definition of such a segment is important because it sets the basis for the search for Innovation Drivers. Discuss who are the key customers of this segment and why. If there is an IA that covers the chosen market segment, data from

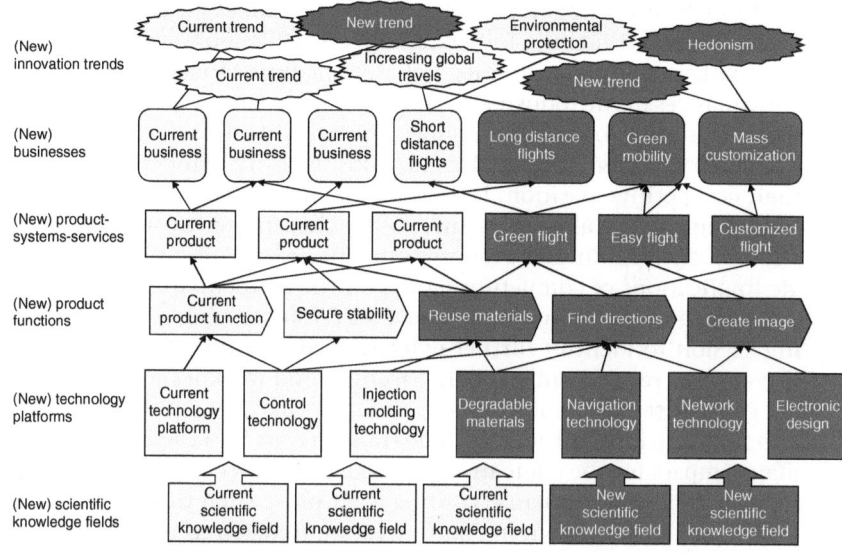

Figure 1.15 Innovation Architecture as a snapshot of a current situation

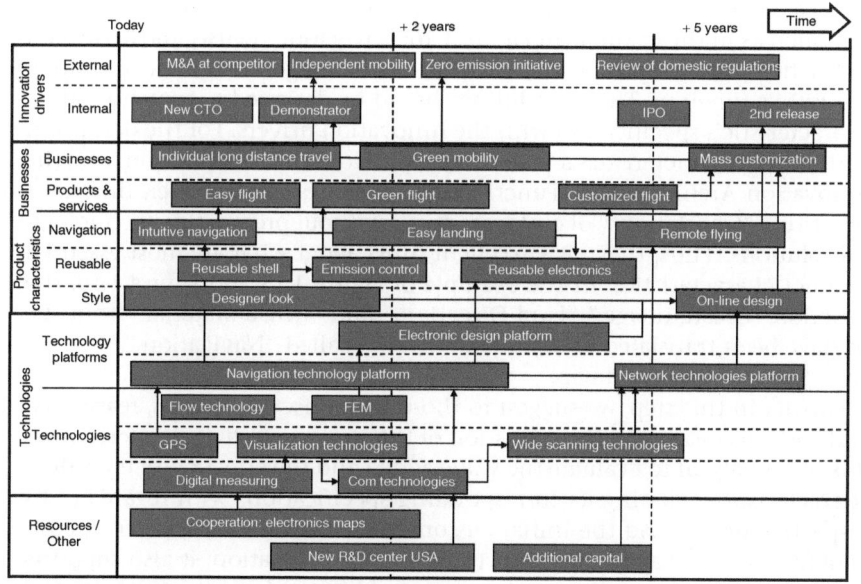

Figure 1.16 Generating a Business Roadmap

the level "Innovation Trend" can be used as sources for external drivers. In general, also include the following aspects while looking for external drivers: market changes, technology changes, competitor moves, cooperations and alliances in the industry, competencies and financials, legal regulations, industry norms, or environmental requirements. Usually competitors represent very strong external drivers. Thus, identify competitors and discuss their competitive position. Doing so, it is important to stay focused on the most important competitors and on the chosen market segment. Consider strength and weaknesses of those competitors, and discuss which the strategic implications of competitor moves for the segment are.

It can be very helpful to perform a SWOT analysis, or a short brainstorming session to identify further external and also internal drivers. Internal drivers may result from company strength and weaknesses and from strategic goals. List those entire drivers on a flipchart, group them meaningful, and finally prioritize them. Include those drivers into the BRM, which are of major impact for the company and for key customers of the segment in consideration, schedule them according their prospected time of occurrence.

In the example shown in Figure 1.16, we transferred two external diverse forms in the IA shown in Figure 1.15: the driver "Zero Emission Initiative" is a derived from the trend "Environmental Protection," and the driver "Independent Mobility" from "Increasing global travel."

Step 2: Defining product characteristics. Product characteristic are to be selected so that they will be able to respond to internal and external Innovation Drivers. We recommend defining three to four key characteristics, such as in the example in Figure 1.16: Navigation, Reuse, and Style, and to further look for specifications within those categories to schedule along the BRM's timeline. This scheduling should be designed to match product characteristic's specification with the innovation drivers. For the definition of those key characteristics it is usually helpful to seek for inspiration from the Innovation Architecture's "Function level." As described earlier, functions constitute "translations" of customer needs without preempting technological solutions. Thus, they give excellent hints about what are most essential product characteristics. In the example of Figure 1.15 and Figure 1.16, the function "Create image," "Find Direction," and "Reuse Material" from the IA have been translated to key characteristics called "Navigation," "Reuse," and "Style" in the Roadmap.

Further in this step, we suggest to thoroughly discussing risks, feasibility, and costs necessary for the realization of the chosen product characteristics. It is often helpful to evaluate the value added and the effect of each product characteristic including its corresponding specifications with regards to its respective driver, and the initial customer need behind the driver. A cost-benefit analysis may be the right tools for this evaluation; it also supports to identify which key characteristics having highest impact on drivers and which have an effect on multiple drivers. Include those key characteristics in the BRM, on the Business level's Product Characteristics layer.

Step 3: Tag the businesses into the BRM according to schedule. Prior to implementation of this third step, intensive discussions about strategic goals of the company's businesses and how/when to achieve them must already be concluded. When including the businesses into the BRM, they should be chosen and scheduled in a way to respond simultaneously the prior discussed strategic business goals, and to the Innovation Drivers. Again, as a source of inspiration for the businesses to serve in the future, the IA will certainly help. In example, the three businesses "Long distance Flights," "Green Mobility," and "Mass Customization" have been adapted from the IA to the BRM as is.

Step 4: Line up the product strategy. This step consists of discussing all possible product strategies that satisfy at the same time the prior defined product characteristics and the defined businesses. If discussions in the earlier Steps 2 and 3 have been done thoroughly, usually the definition of the product strategy can be regarded as a logical consequence. For the discussion about the product design, it is often helpful to write down possible product concepts in a simple "brochure like" language and make to simple sketches on a flipchart for innovative product designs. These two simple methods will enhance the discussions when performing this level as all involved people will have a more concrete image of the products in discussion. Further aspects to include into the discussion might be: the choice about fundamental product strategy decisions such as "high-end" or "low-end" product variants, product family concepts, and platform strategies, the deadlines for prototypes, timelines for market introductions and for new product generation releases, etc. In example, the three products "Green Flight," "Easy Flight," and "Customized Flight" have been adapted from the IA to the BRM as is.

Step 5: Schedule Technological solutions. This step comprises two major tasks: first, identifying technological solutions, and second, evaluating these solutions with regards to their contribution to realize all the product characteristics decided upon in Step 2.

The ultimate goal is to create transparency about potential technology developments or acquisition activities related to the decisions in the BRM made so far. The identification of potential technological solutions goes thus hand in hand with their assessment. The most important questions to clarify during this assessment are: how suitable are the potential technologies for the realization of given product characteristics, how is the company's own level of knowledge about them and how is the general level of knowledge about them. These questions give an insight about the effort and time necessary to accumulate or in-source the technologies in question. Further discussions about the strategic importance of the technology with regards to the chosen business strategy are also recommended. Such discussion gives hints about the competitive advantage related to a given technology, for example its potential to support a differentiation- or cost-leadership strategy. Experience shows that this level of the BRM is most effectively processed if only those technologies are included that are new to the company. Existing and already mastered technologies should not be specified at this moment in time. Again, the Roadmap can benefit from the insights already elaborated in the IA, they can be taken as an initial position for all the discussions necessary in this step.

Step 6: Forming Technology Platforms. The formation of technology platforms suggests analyzing once again the chosen technologies from the previous step in order to group them into meaningful strategic entities. The appointment of technology platforms should be done in alignment with the strategic discussions from the previous step. Basically speaking, technology platforms should be clustered according to the ambition of the company to build up distinct competencies in a certain technology field, and to be able to take advantage of the platform's synergetic effects over multiple business units.

Also, the thoughts and insight gained at the elaboration of the platforms in the IA are usually a great help. The experience shows that the discussion about the content of possible technology platforms in this step should include both the new technologies to build up – those were discussed in the previous step – as well as already existing and mastered technologies. It will give the most complete image about possible platforms and their impact. It is, however, still not recommended to include existing technologies in the BRM, rather the detailed design of platforms should be subject to separated discussions.

Most often, the identification and evaluation of technologies performed in Step 5 and the decisions about the technology platforms in this current Step 6 is an iterative and somewhat time consuming process.

Step 7: Planning implications for resources. Usually, planning activities in the BRM over a mid- to long-term period will require major efforts at implementation. Thus, we also suggest including in the BRM the implications for the company's unit's resources. Subjects to be discussed are typically: addition capital or higher budgets that will be necessary, strategic alliances in R&D, or in logistics, or the necessity to set up new project- and organizational structures.

Final step: Only after all the levels of the Roadmaps are properly elaborated do we suggest beginning the linking of items of each level by connectors and arrows showing their direct relationships over time. Doing so means to definitely set and align market related milestone and development related timelines. This last step of linking items usually brings great value to the overall accuracy of BRM's planning activities because all items are usually critically reviewed and considered in an overall planning context including all goals, activities, and timelines and handovers between levels.

The BRM elaborated according to the above-described steps represents the technology- and innovation-strategic plan. This strategic plan should not be confused with the more operational project planning. Once the BRM is completed concrete technology-, product-, business-, and market development projects have to be derived, scheduled, and implemented according to state of the art project management.

Practical applications in workshops

This section describes the major challenges encountered when practically applying the BRM – typically in workshops – for strategic planning in a company.

Getting started: For those who may want to start with a single application of the BRM, a good occasion is the evaluation and planning of a specific business opportunity. It is however more beneficial to develop and review a Roadmap on a regular basis, for example for the periodical strategy update in every business unit.

Being prepared: Preparations before the first introduction of the BRM should include the following points: the clarification and communication of the intention and the ultimate goal of introducing the BRM. This task is best insured by one of the top managers involved in strategic planning who has a clear self-interest in the BRM's implementation.

Furthermore, the object of planning should be well defined and confined, so that it is clear where to recruit the people from to involve in the BRM process. Finally, the timeline necessary to develop the roadmap including development procedure's main steps should be well communicated to all participants.

Choosing the implementation method, and the people to involve: Generally speaking, there are two different alternative implementation

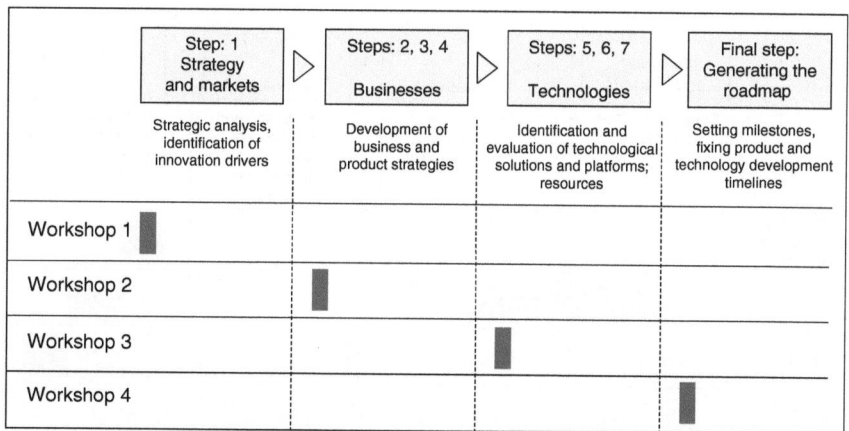

Figure 1.17   Elaborating a BRM in practice – the detailed workshop schedule

methods that proved to be most effective. They differ in the necessary time and depth when to develop the BRM.

The first alternative showed best results in the case companies introduced the BRM for the first time as a planning tool. It is the detailed and more time-consuming workshop schedule shown in Figure 1.17. It allows a diligent elaboration of the BRM in four different workshops, each of them developing one of the three main levels of the BRM, and one additional closing workshop in which the actual Roadmap is generated. For each workshop, half a day should be reserved for an interdisciplinary group of a maximum of ten people. It is emphasized here that it is most critical to compose this group as an interdisciplinary team of people with different functional backgrounds. Represented company functions should include R&D, marketing, sales, and finances. Such a group elaborating a TRM in an intersubjective way will realize the most accurate planning results. Best results can be expected if the composition of the team is the same for all four workshops.

The four workshops should each be prepared, moderated, and revised by an extra moderator with good communication skills and experience about the BRM elaboration process.

All in all, the detailed workshop schedule for developing a BRM will take about two to three months, depending on the time between the workshops. Experience shows that most of this time is required between Workshop 1 and 2. This is due to the fact that usually the discussions about the basic strategy in Workshop 1 will raise new open issues that had not been covered by the initial strategic analysis available at the workshop. The time till Workshop 2 is thus the opportunity to do this additional strategic analysis, as well as to prepare thoroughly for the development of the business and product strategy in Workshop 2. More or less one month between each workshop should be given.

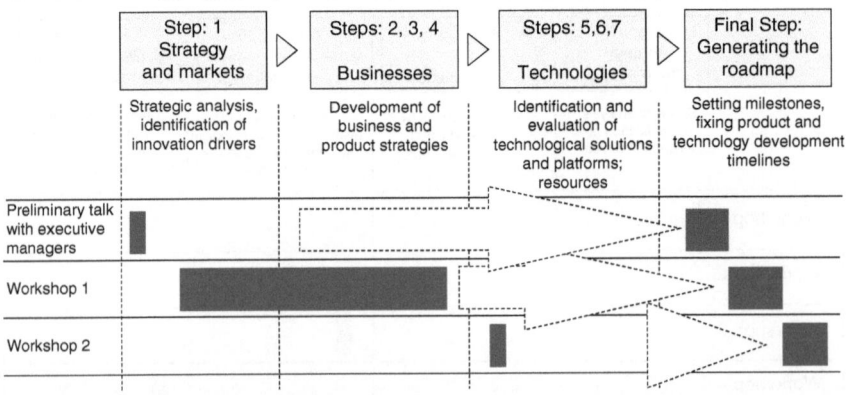

Figure 1.18 Elaborating a BRM in practice – the time compressed schedule

Figure 1.18 shows the second alternative of scheduling and running through workshops. This procedure will be more rapid to develop a Roadmap. It is recommended if a company has already gathered some experience with the BRM process, and in order to update an existing BRM.

Timesaving in this compressed schedule is realized above all by preliminary talks with executive management. They gather strategy related information in a most direct and complete way before starting with the BRM process as such. Usually these talks are done by concerted interviews; the corresponding information is collected and reviewed by the moderator. This way, usually all the information necessary to perform Steps 1 to 4 in one single workshop can be brought together. At the same time, the generation of the actual BRM is done as the information and decisions are available, meaning the actual BRM is generated ongoing and not all together in a final closing workshop. This is symbolized in Figure 1.18 by the arrows pulled through to the Final Step. This way, the rapid schedule enables to deliver an update of an existing BRM within about one month's time.

Keeping the BRM alive: Once the BRM has been elaborated it is important to keep it alive and up to date. Thus, at least two meetings per year should be scheduled where the basic assumptions and decisions of the Roadmap are critically reviewed and updated. A reliable method to keep the Roadmap alive is to establish it not only as a planning tool, but also as a communication tool. For example, in one case we established the BRM – together with the IA – as the two major communication and reporting tools for the periodical R&D managers meeting with the company's CTO. Each BU's R&D manager first reported their overall plans by using the BRM to show the current planning status in their BU, and then broke down this image to the different operational projects. This approach enabled the CTO to not only review the projects within the various BUs, but also to more easily recognize and initiate measures to align and synchronize strategic R&D activities across BUs to foster synergies and leverage resource efficiency.

Business Model

The term "model" has its roots in the Latin word "modulus" with translations like "measure," "scale," and "pattern". It is used in science and technology – and even in philosophy – with hardly perceivable diversity. However one meaning seems to prevail in many cases: "model" to refer to a physical or conceptual construct of a particular reality. An example in the first case is a tangible reconstruction of a bridge, downscaled by a factor of 100. A conceptual model, for example, is Niels Bohr's atomic model, the "Bohr Model". The models in both cases represent simplifications of the reality in such a way that on the one side essential characteristics of the reality are still present. On the other side they allow investigating the model behavior in order to come up with results which permit conclusions to be drawn for the reality.

In the first case the model of the bridge can be exposed to varying wind conditions in a wind tunnel and thus deliver results on how to avoid design parameters – so-called resonance frequencies – which unavoidably would cause the fate of the Tacoma Narrows Bridge on November 7, 1940. In the second case the most primitive concept of the Bohr Model – a positively charged nucleus surrounded by negatively charged electrons traveling on an orbit around the nucleus – allowed nuclear physical calculations which led to a confirmation of the so-called Rutherford formula which had been known before only experimentally.

In a similar sense BMs are being created. They follow the purpose to visualize graphically a company where the essential considered characteristics are still preserved. The almost infinite number of existing and potential BMs varies primarily in the choice of these characteristics.

The interest in investigating BMs can be exemplified, for example, with reference to an MIT Sloan School publication which analyzes the BMs of all 10,970 publicly traded companies in the US economy from 1998–2002 (Malone et al., 2006).

For the purpose of this article the BM suggested by Osterwalder (2004) has been selected. In a first step four main areas which represent the essential topics of managerial concern are identified (Osterwalder, 2004: 42):

- Product: What business the company is in, the products, and the value propositions offered to the market.
- Customer interface: Who the company's target customers are, how it delivers their products and services, and how it builds a strong relationship with them.
- Infrastructure management: How the company efficiently performs infrastructural and logistical issues, with whom, and as what kind of network enterprise.
- Financial aspects: What the revenue model is, cost structure, and the BM's sustainability are.

In a second step these main areas have been split into nine BM building blocks which reflect a synthesis of the BM literature reviewed. They are presented in Figure 1.19.

Keeping the basic structure and their underlying reflections of the BM as suggested by Osterwalder (2004) in mind, for the purpose of this publication a generic structure has been developed which is presented in Figure 1.20. On the one side it contains all the nine building blocks according to Figure 1.19. On the other side, an additional four elements have been built into the BM: The first element "competition" visualizes the company in its competitive environment. Next is "Technologies": This element allows expressing essential core competences of strategic significance for a technology driven company. Third is "Profitability": This provides information on the overall financial performance. And, finally, "Growth

Main areas	Building blocks of business model	Description
Product	Value proposition	A value proposition is an overall view of the company's bundle of products and services that are of value for the customer.
Customer interface	Target customer	The target customer is a segment of customers a company wants to offer value to.
	Distribution channel	A distribution channel is a means of getting in touch with the customer.
	Relationship	The relationship describes the kind of link a company establishes between the customer and itself.
Infrastructure management	Value configuration	The value configuration describes the arrangement of activities and resources that are necessary to create value for the customer.
	Capability	The capability is the ability to execute a repeatable pattern of actions that is necessary in order to create value for the customer.
	Partnership	A partnership is a voluntarily initiated cooperative agreement between two or more companies in order to create value for the customer.
Financial aspects	Cost structure	The cost structure is the representation in money of all the means employed in the business model.
	Revenue model	The revenue model describes the way a company makes money through a variety of revenue flows.

Figure 1.19   The nine business model building blocks
Source: Osterwalder, 2004: 43.

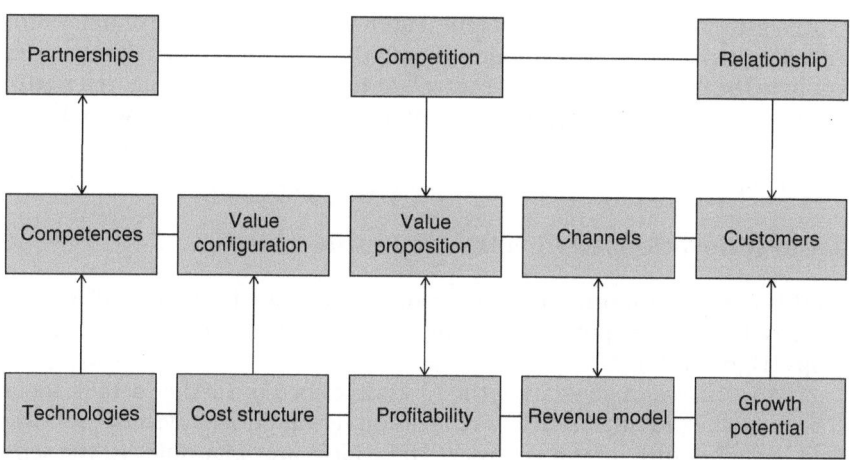

Figure 1.20   Generic structure of a business model
Source: Osterwalder, 2004: 43.

Potential": This element illustrates the prospective business development in coming years.

A practical example of this BM is illustrated in Figure 1.21. It concerns a company by the name of Cerberus AG, the pioneer and world leading

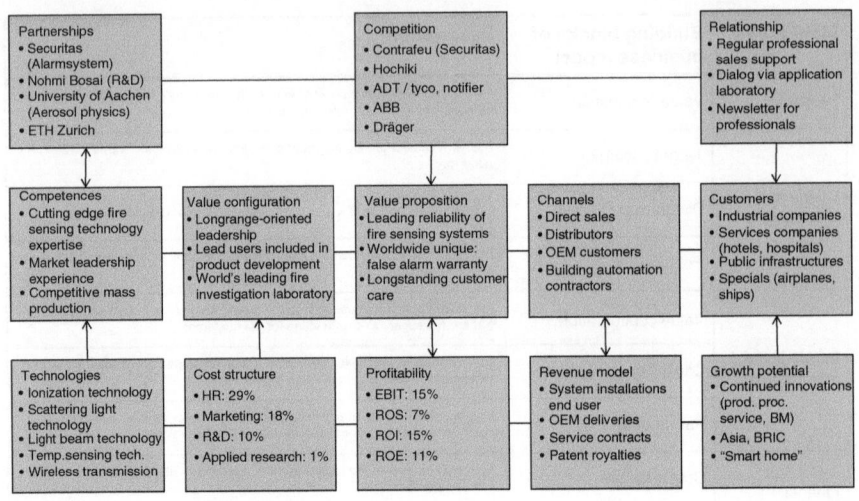

Figure 1.21 Business Model example: Cerberus AG

enterprise in the field of early warning fire detection founded in 1941. Since 1998 this company belongs to Siemens and constitutes the core business of the Buildings Technologies Division of Siemens.

In addition to the factual content, Figure 1.21 mirrors the exemplary role of a BM presentation: To provide a descriptive and comprehensive overview on how the company is "functioning." At the same time it reflects a significant principle of management communication: providing essential overview information on a single sheet of A4 paper, rather than many pages of a long report.

From Innovation Architecture to Business Model

In this last part a summarizing procedure is described, which leads systematically from developing the IA, the ISM, and the BRM to the BM in five steps (Figure 1.22):

Step 1 consists of generating the IA as described in further details above. The outcome provides a variety of ideas on new products, systems, services, and corresponding new businesses. These ideas are "plausible" in the sense that they represent perspicuous matches of trends, needs, available knowledge, and technologies. Being plausible does not yet mean being realistic in a business sense. Therefore the "plausible" business ideas have to be analyzed in order to come up with "realistic" options for new businesses to be subjected to a business decision.

This is the purpose of step 2. Lets assume that the designed IA has come up with four prospective new businesses A-D (Figure 1.22). In order to

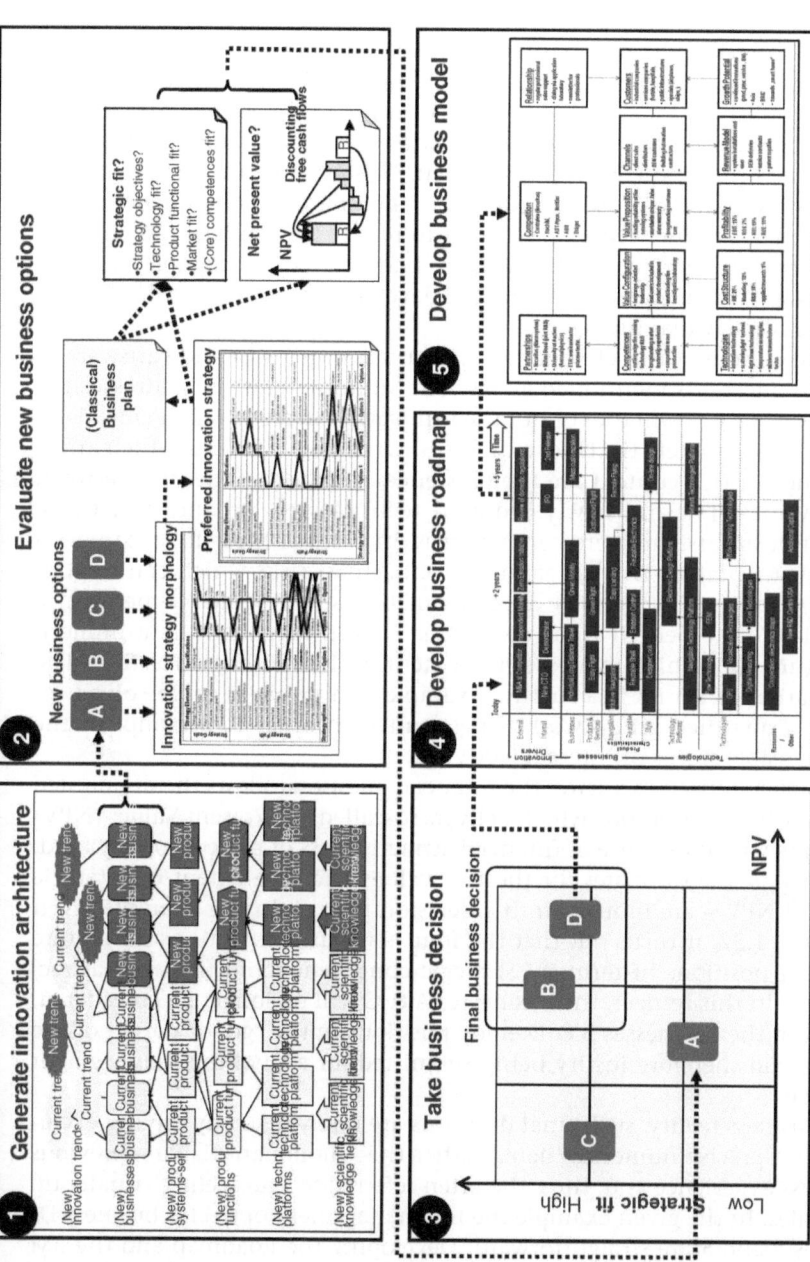

Figure 1.22 From Innovation Architecture to business model in five steps

take a selective entrepreneurial decision, a strategic and a financial question have to be answered. To this end concrete ideas have to be developed on how to realize the new business options. This requires first building up appropriate strategies for each one of the prospective businesses. This can be accomplished, for example, by way of using the presented SM. Its outcome consists of a limited number of strategy options, among which the "Preferred Innovation Strategy" is chosen. This provides the basis for a classical Business Plan (BP), which elaborates the financial consequences of all operational activities required in order to realize a given business strategy. It usually closes with a detailed business budget over, for example, the next five years. Within this time frame the development of the Free Cash Flow (FCF) is of particular interest.

The outcome of the BP together with the SM helps provide answers to the two questions mentioned above. The strategic question addresses the still open issue of whether or not the innovation strategy under consideration does adequately fit the overall business strategy. Criteria which are of influence in this context are the existence of realistic synergies between already mastered and newly required technologies, product functions, market positions, and core competences. In cases of high discrepancies a risk/return analysis may reveal limits of business risks which can responsibly be taken. The various criteria can be used in order to perform a so-called Value Benefit Analysis (VBA). This in turn delivers a numerical Figure 1.22 which expresses the extent of the "Strategic Fit (SF)." The elaboration of a VBA is a further example of a problem to be solved for which a qualified group work is best suited in order to accomplish the indispensible status of intersubjectivity.

A usual approach to answer the financial question follows the discounted "Free Cash Flow Method" which delivers so-called Net Present Values (NPV) as a widely used measure of financial attractiveness of projects of any kind.

In step 3, the estimates for the two values – for the extent of "Stategic Fit" and NPV – are brought into a decision matrix. In the example given in Figure 1.22, it turns out that the four new business options have quite different positions in terms of strategic congruency and financial attractiveness. In this respect, the businesses A & C will no longer be considered. However, the businesses B & D enjoy positions in the "good" corner of the matrix and therefore justify being submitted for a final top management decision.

In business reality, such final decisions are rarely taken by relying exclusively on precise numerical data. Rather the – nonquantifiable – previous business experience and, thus, the often referred to "gut feeling" finally tip the scales. In the given example the top management opted for business D.

Steps 4 and 5 are straightforward: Developing the Roadmap and the BM follows the final decision – in our case for Business D – according to the guidelines given in Chapters 4 & 5.

Managerial implications

- Developing innovation strategies for Strategic Business Units (SBUs) represents a managerial task which requires the integration of all entrepreneurial functions – without exception. Usually a corresponding process is faced with intra-organizational cultural challenges, since each function, such as marketing, finance, production, and R&D is focused on specific views of products and services, which at first are impeding the finding of a consensus.
- In order to cope with these inherent challenges, five managerial tools can be used which are explicitly suited for working in groups in which all functions are represented.
- The first tool – the Innovation Architecture (IA) – constitutes as systemic as well as systematic procedure, which allows merging the function related perspectives such as innovation fields, new societal and marketing trends, emerging customer needs, new technologies, newly relevant scientific knowledge in such a way, which results in – at first sight – plausible ideas for new products, new services, and new businesses.
- The second tool – the Innovation Strategy Morphology (ISM) – allows in a first step to discuss and display all conceivable strategic elements and there specifications. For this discussion it is recommended to distinguish between "strategic goals" and "strategic paths" in order to clearly differentiate between strategic elements which are related to the next higher hierarchical level ("strategic goals") and those which represent the "space of action" within which free decisions can be taken by the SBUs. In a subsequent step, group discussions can again take place on meaningful combinations of individual specifications of all strategic elements. Such combinations represent concrete innovation strategy options which have to be investigated further with respect to their business attractiveness.
- The result of this investigation is displayed in the third tool – The Business Decision Matrix (BDM). It allows positioning each innovation strategy of each business option developed in the IA according to values of the two dimensions: Strategic Fit and NPV. These two values result from traditional business plan analysis. In this matrix, final management decisions can be taken from business options located in the matrix corner with high values for both variables.
- Having decided on a new business, it is required to consider its timely realization. The appropriate tool is presented as the fourth tool – the Business Roadmap (BRM). This tool displays the carefully reflected timely synchronization of all individual business components such as financial resources, newly required competences, technologies, and products in order to enter the selected markets on time.
- The fifth tool – the Business Model (BM) – represents a quite popular way to explain a business on one singular scheme.

- In practice, the application of these tools is quite straightforward and does not require lengthy explanations. Moreover they fulfill an unwritten principle of management communication: to present business ideas in a self-explanatory form on one sheet of A4 paper.
- In addition, using these tools facilitates comparing and reporting different business options across divisional boundaries.

Notes

1. Daily Yomiuri Shimbun October 13, 2007.
2. With the help of a web camera with a resolution of 300,000 pixels or more, the computer recognizes three points of the eye: the inner corner of the eye, the inner extremity of the eyebrow and the center of the pupil. For adjusting the system, the user has to look at one of the 15 on-screen buttons for one second and blink and to repeat this procedure for all 15 on-screen buttons.

   Having finished the adjustment process, the program is ready to be used. In operation the 15 on-screen buttons refer to 10 hiragana characters – a, ka, sa, ta, na, ha, ka, ya, ra, wa – and five function keys such as the "enter" key. When a user wants to input "u" in hiragana, for example, he gazes at the "a" button for one second and blinks. The computer then shows the "a" column of the hiragana syllabary – a, e, i, o, u. Looking now at the "u" button for one second and blink, the "u" character will be displayed. Hiragana characters are finally converted into Chinese characters by using a dedicated on-screen function key.
3. http://www.newkast.or.jp/english/projects/pro_nakamura.html.

References

Malone, T. W., Weill, P., Lai, R. K., D'Urso, V. T., Herman, G., Apel, Th. G., and Woerner, St. L. (2006). *Do Some Business Models Perform Better than Others?*. Working Paper: MIT Sloan School of Management.

Osterwalder, A. (2004). *The Business Model Ontology – A Proposition in a Design Science Approach*. Doctoral Dissertation: Ecole des Hautes Etudes Commerciales de l'Université de Lausanne.

2

Organization of International Market Introduction: Can Cooperation between Central Units and Local Product Management Influence Success?

Antje Baumgarten, Cornelius Herstatt, and Claudia Fantapié Altobelli

Introduction

Developing consumer winning new products is quite a challenge for every company, but even more for a multi national corporation (MNC). To take advantage of synergies and cut costs they need to develop standardized new products that can be launched homogeneously in as many markets as possible. Executing its "path to growth programs" Unilever has reduced its number of brands tremendously and wants to focus solely on so called "leading brands" – strong brands which can be offered globally (Unilever, 2002). Following this strategy of developing standardized global products the organization of international market introduction is more and more centralized: for example, Unilever, Masterfoods, and GlaxoSmithKline have founded "skunk works." These are separate units, usually in a different location, where a dedicated team is working centrally on innovation for the European or global market (Murphy, 2003). Once the product and the launch strategy are finalized, the central unit hands it over to the subsidiaries.

But in reality, consumers are not as homogenous as these central and standardized approaches wish for. Culture, language, habits, and needs are still pretty diverse across local markets. Quite often international market introductions fail because diverse local consumer needs have not been taken into account sufficiently. Since the knowledge about local consumer needs lies within the subsidiaries, it is rather questionable if a central management of developing and introducing a new product without or just a late local subsidiary involvement can fulfill local consumer needs sufficiently.

But if the new product does not meet consumer needs it is doomed to fail. In addition, a very late or even no involvement of local managers during the process of an international market introduction can lead to acceptance and coordination problems. In the worst case they can simply reject launching the new product in their market.

Assuming that cooperation between the central unit and subsidiaries would positively influence consumer need based international market introduction of a new product and its local acceptance, the research objective of this study is to investigate whether cooperation with local subsidiaries can improve the success of centrally organized international market introductions.

First we describe the process of international market introduction and look more specifically into the definition and impact of cooperation. Based on a set of hypotheses drawn from recent research we build a conceptual research framework, which is validated by a quantitative survey. Based on the survey results we will draw conclusions and give detailed recommendations on how to improve cooperation in international market introduction.

Research framework

The objective of an international market introduction is a successful launch of a new product. Thus, the international market introduction process summarizes all strategic and tactical marketing planning and decision-making processes describing how these processes have to be broken down into consumer, competitor, and company related tasks (Call, 1997; Lach, 2001). The company decision to develop and launch a new product marks the starting point of this process. The end point is reached when all objectives that have been set specifically for this market introduction are achieved (Figure 2.1).

International market introductions usually aim to launch a new product globally in markets where a company is already present. Assuming that a standardized and centralized approach is taken, it becomes evident that the implementation of this process is a complex task. Just one single unit of the MNC can hardly fulfill it. Therefore roles and responsibilities are usually shared between central and local units. The central unit is responsible for initiation of the process, steering, and control. Usually this central unit is organized as European or global product management, innovation department or lead country. At a local unit level the local product management of the subsidiary usually assumes responsibility. But if roles and responsibilities are shared, a fair amount of coordination and cooperation is required.

In general the term "cooperation" describes a common fulfillment of a task within a company. It also describes the scope and intensity of this common fulfillment (Schreyögg, 1999; Zentes, 1992). In the context of innovation management cooperation is the common term to describe collaboration between different functional units, for example Pinto and Pinto

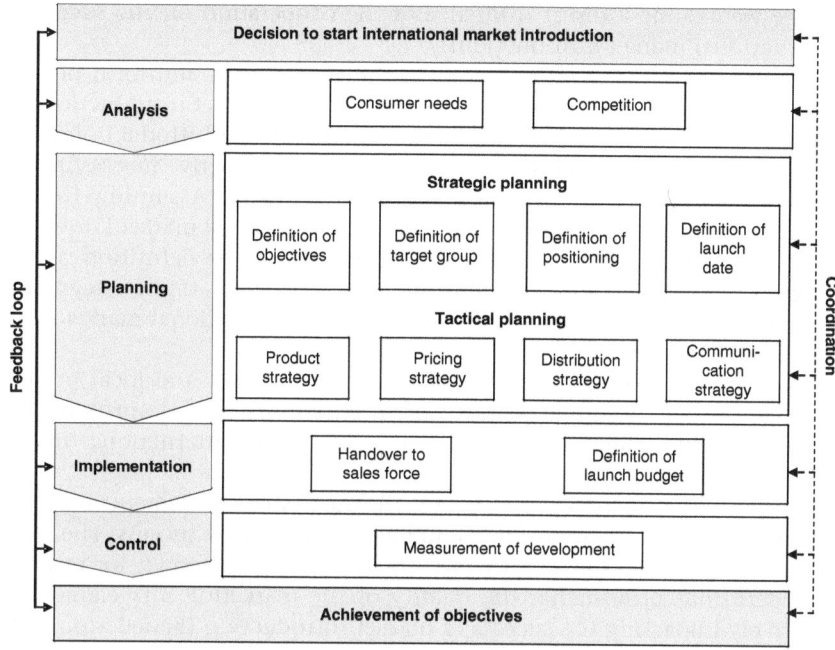

Figure 2.1   International market introduction process overview
Source: Sauter, 2001; Meffert and Bolz, 1998.

use cooperation to describe collaboration between R&D and Marketing (Pinto and Pinto, 1990). Therefore, in the context of international market introductions cooperation between a central and a local unit shall describe the scope and intensity with which each task of the process is fulfilled by both parties together.

Research on cooperation in international market introduction to date is quite limited. Chryssochoidis and Wong have confirmed that a high degree of cooperation is positively influencing timeliness of international market introductions (Chryssochoidis and Wong, 1998). Further research has shown that, in particular, communication between central and local units is positively affecting new product acceptance by subsidiaries (Boutellier and Lach, 2000; Goshal and Bartlett, 1988). Thus, we can assume that cooperation between the central unit and local product management could have a positive effect on the success of international market introductions. All local consumer information could be considered early in the development. This would lead to improved timeliness, better acceptance of the new product by the subsidiary, and probably higher acceptance by local consumers.

Hence we assume a direct influence of the cooperation on the success of international market introductions:

Hypothesis 1: Cooperation between the central unit and local product management in terms of all tasks of international market introduction can positively influence the success of international market introductions.

But cooperation could also influence success indirectly. Successful new products are based on sufficient market knowledge. Assuming that an increase of cooperation would lead to an increase of local market knowledge by the central unit, this could lead to better target group definition, a more focused positioning, and a more influential communication strategy, thus to a better overall quality of all elements of the international market introduction. Hence:

Hypothesis 2: Cooperation between the central unit and local product management in terms of all tasks of international market introductions can positively influence the quality of elements of international market introductions.

An indirect influence on success via the quality of the elements would be given only, if the quality itself would have an impact on success. The influence of quality has been confirmed by a number of surveys, for example Benedetto has proven that the quality of the marketing mix elements is positively impacting the success of market introduction (Benedetto, 1999). In this study we look specifically at the impact of the quality of the elements of market introduction in an international context. Hence:

Hypothesis 3: The higher the quality of elements of international market introductions, the higher the success of international market introductions.

A key problem of market introductions are changes made late in the process. When the new product is designed, 90 percent of the costs required later are agreed in this early phase. Therefore it is crucial that all information is available and considered early on in the process. Any changes that have to be made later on can lead to a cost increase and a delay of the launch. In addition, if these required changes have to be made under time pressure, this may lead to half-hearted solutions (Boutellier, Corsten, and Lach, 1997). Especially when these changes have to be made after launch which can be very expensive. Examples from the automotive industry show that calling back cars from the market requires substantial payments of damages, can lead to losses in image ratings and, more severely, lack of turnover (Boutellier and Lach, 2000). An intense cooperation with local product management could make sure that all required local information is taken into account early on which may prevent costly changes later on in the process. Hence:

Hypothesis 4: Cooperation between the central unit and local product management in terms of all tasks of international market introductions can reduce the degree of changes made after launch.

Thus cooperation could influence success of international market introduction also indirectly via changes made after launch, but only if these

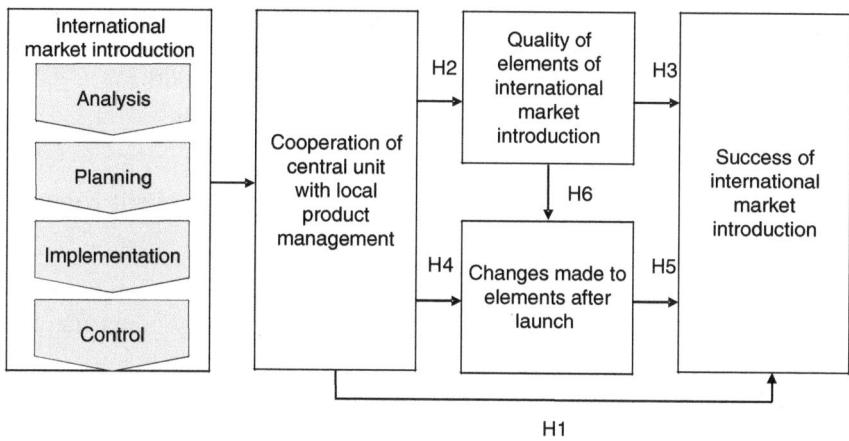

Figure 2.2　Research framework

changes themselves would influence success negatively. For national market introductions there is already evidence that successful new products have shown less problems during production and after launch (Rothwell et al., 1974; Rothwell, 1985). Hence we propose the following for international market introductions:

Hypothesis 5: The fewer changes that are made to the elements of international market introductions after the launch, the higher likelihood of success in international market introductions.

Yet the degree of changes made after the launch is probably influenced by the quality of the elements. A high product quality will lead to a higher acceptance rate by consumers and won't require any changes and if the communication mix is working, again no changes will be necessary. Hence:

Hypothesis 6: The higher the quality of the elements of international market introductions, the fewer changes are made after launch.

These hypotheses represent our research framework enabling us to validate the impact of cooperation between the central unit and local product management on the success of international market introduction (Figure 2.2). Since every hypothesis needs to be examined for every task of the process we have formulated sub-hypothesis respectively (see appendix).

Research methodology

Respondents and industry

To test our research framework we used a key informant sample of local product managers at subsidiaries of MNCs in Germany, Austria, and Switzerland. We have chosen MNCs from the consumer goods industry only, first to keep the sample homogenous to enable clear managerial recommendations (Schröder, 1994) and secondly because this industry has been chosen for research on NPD success factors just rarely (Haake et al., 2000).

The sample selection was based on the following criteria:

subsidiaries of MNC of the consumer goods industry
with a minimum turnover of 50 Mio. € to ensure a sufficient company size, and
owning at least 3 foreign subsidiaries to qualify as a MNC.

In addition every subsidiary had to have its own local product management that is involved in international market introductions.

Addresses were drawn from the company databases of Hoppenstedt and Kompass, and member lists from the Swiss Society of Marketing, Society of Companies Selling Cosmetics and Promarca. In total 211 subsidiaries meeting these criteria were identified. Then every subsidiary was called to identify the name of its marketing director.

Data collection

In parallel a mail survey questionnaire was developed and pretested. Since the total number of subsidiaries identified was quite low, we decided to mail three questionnaires to each subsidiary in order to boost sample size. The questionnaires were mailed to the local marketing director in March 2004 with a request to distribute it to his local product managers. Every local product manager was asked to select the latest international market introduction that he has been working on, but which has been initiated and managed by a central unit (last incident method). In total, 51 questionnaires were returned, each of them representing a distinct international market introduction.

Measures

Based on the research framework we first asked in general whether cooperation between the central unit and local product management took place or not for each task of the process. If they were cooperating, we let the local product manager evaluate intensity and quality of the cooperation (Lawrence and Lorsch, 1967). Similar to Langerak, we used a five-point rating scale to measure the intensity of cooperation ranging from: one being very low to five equalling very high. We also measure the quality of cooperation ranging from one being very bad to five being very good (Langerak et al., 1997). The quality of elements was measured from one being very bad to five being very good. Degree of changes made after launch was measured with one being very small to five being very extensive. Finally, success of the international market introduction was also evaluated on a five-point rating scale ranging from one equalling a total failure to five being a big success.

Results

Cooperation with local product management as direct success factor

First we looked at every single task of the process and divided the sample in a group where a cooperation took place and a group where no cooperation was practiced. A F-Test was performed to test both group for significant difference

Table 2.1 Correlation of quality of cooperation and success of international market introductions

Hypothesis	Task	Correlation coefficient, r (Quality of cooperation and success)	Hypothesis confirmed (✓) rejected (x)
Analysis			
H1a	Analysis of consumer needs	0.487**	✓
H1b	Analysis of competition	0.406*	✓
Strategic planning			
H1c	Definition of objectives	0.496**	✓
H1d	Definition of target group	0.470**	✓
H1e	Definition of positioning	0.490**	✓
H1f	Definition of launch date	0.396**	✓
Tactical planning			
H1g	Definition of product concept	0.596**	✓
H1h	Definition of brand name	n.s.	x
H1i	Design of packaging	n.s.	x
H1j	Definition of product range	n.s.	x
H1k	Definition of pricing strategy	0.490**	✓
H1l	Definition of communication strategy	0.383*	✓
H1m	Definition of distribution strategy	n.s.	x
Implementation			
H1n	Hand over to sales force	n.s.	x
H1o	Definition of launch budget	n.s.	x
Control			
H1p	Measurement of development	0.315*	✓

* Significance at the 0.05 level.
** Significance at the 0.01 level.

in terms of success of the international market introduction. Results showed no significant differences, indicating that cooperation as such is not critical for the ultimate success. This is probably caused by measuring cooperation on a general level, rather than on its individual dimensions, for example the number of visits made or number of e-mails exchanged. Even though where a way of cooperation existed, for example some e-mails were exchanged, the actual quality might have been quite low. Thus we postulate that intensity and quality of cooperation significantly influence market success.

When cooperation took place we asked local product managers to evaluate intensity and quality of cooperation. First we looked at the independence of both dimensions via correlation analysis. The evaluation of intensity and quality of cooperation showed a significantly positive correlation with the exception of definition of brand name, product range, and distribution strategy. Thus quality of cooperation is evaluated as very good if its intensity was very high. Therefore we were focussing on quality of cooperation only for testing our hypotheses.

Hypothesis 1 predicts a direct influence of cooperation on success. To validate this hypothesis the degree of correlation between quality of cooperation and success of market introduction was measured. Correlation shows significantly positive results for the majority of tasks (see Table 2.1). Exceptions are definition of brand name, design of packaging, definition of product range, definition of distribution mix, hand over to sales force, and definition of launch budget. Hypothesis 1 can therefore be partly confirmed.

These results show that quality of cooperation has a positive impact on success for almost all tasks of international market introductions, but especially for tasks in the early phases of analysis and planning. In the later phase of tactical planning, and especially in the phase of implementation, there is almost no influence. Here local product management can probably decide for itself about tasks and elements. Thus, these are already tailored to local needs and cooperation with the central unit is not critical for success anymore. Yet in the phase of control, cooperation becomes crucial again. Here the central and local unit need to analyze together the new product performance in the market and have to agree on adjustments needed.

Cooperation with local product management as indirect success factor

In addition to the direct influence of cooperation on success we also wanted to test the indirect influence. Based on Hypothesis 2 we looked at the correlation of quality of cooperation on selected tasks and the quality of the corresponding elements of international market introductions. As results in Table 2.2 show, Hypothesis 2 can be partly confirmed. The analysis confirms that cooperation with local product management when defining the positioning has a significantly positive influence on the positioning quality.

Table 2.2   Correlation of quality of cooperation and quality of elements

Hypothesis	Task/Element	Correlation coefficient, r (Quality of cooperation and quality of element)	Hypothesis confirmed (✓) rejected (x)
H2a	Definition of target group	n.s.	x
H2b	Definition of positioning	0.456**	✓
H2c	Definition of product strategy	n.s.	x
H2d	Definition of pricing strategy	0.405*	✓
H2e	Definition of communication strategy	0.328*	✓
H2f	Definition of distribution strategy	n.s.	x

* Significance at the 0.05 level.
** Significance at the 0.01 level.

Table 2.3   Correlation of quality of elements and success of international market introductions

Hypothesis	Element	Correlation coefficient, r (Quality of element and success of market introduction)	Hypothesis confirmed (✓) rejected (x)
H3a	Definition of target group	0.536**	✓
H3b	Definition of positioning	0.527**	✓
H3c	Definition of product strategy	0.438**	✓
H3d	Definition of pricing strategy	0.448**	✓
H3e	Definition of communication strategy	0.492**	✓
H3f	Definition of distribution strategy	0.529**	✓

* Significance at the 0.05 level.
** Significance at the 0.01 level.

The same is true for pricing definition and communication strategy. Thus, the quality of these strategies can be improved when the central and local unit are cooperating. With regard to target group definition, we assume that target groups are usually fixed – meaning there is a specific definition that is not changing for every new market introduction. It is just a question of selecting the right target group, which can probably be done by the central unit without intense cooperation. The same is true for the distribution strategy. In the consumer goods industry, distribution is handled via retailers, but the retailer landscaped is quite fixed. Thus the distribution strategy is basically fixed and can often be decided by the local product management only – cooperation between central and local units is therefore not required.

Again, any indirect influence of cooperation on success via the quality of the elements would only exist if the quality itself had any influence on success – as stated in Hypothesis 3. As shown in Table 2.3 we can report a significantly positive correlation between the quality of elements of international market introductions and success. Thus Hypothesis 3 is confirmed: The higher the quality of elements, the bigger the success of an international market introduction.

Apart from the indirect influence of cooperation on success via the quality of elements, we proposed another indirect influence via changes made after the launch. Based on Hypothesis 4 we assessed the influence of cooperation with local product management on the degree of changes made after the launch. As the results of the correlation analysis show, a significant correlation between cooperation and degree of changes could not be measured for any of the activities (Table 2.4). Thus Hypothesis 4 is being rejected. However, an analysis of the means and standard deviation showed that the degree of changes made after launch were only quite low.

Hypothesis 5 proposed that the fewer changes that made after launch, the higher the success of international market introductions. We again applied

Table 2.4 Correlation of cooperation and degree of changes made after launch

Hypothesis	Task/Element	Correlation coefficient, r (Cooperation and degree of changes made after launch)	Hypothesis confirmed (✓) rejected (x)
H4a	Definition of target group	n.s.	x
H4b	Definition of positioning	n.s	x
H4c	Definition of product strategy	n.s.	x
H4d	Definition of pricing strategy	n.s.	x
H4e	Definition of communication strategy	n.s.	x
H4f	Definition of distribution strategy	n.s.	x

* Significance at the 0.05 level.
** Significance at the 0.01 level.

correlation analysis to test the influence of the changes made on success (Table 2.5). We can confirm a significant negative correlation only for the pricing strategy, concluding that the fewer the changes made to pricing after launch, the higher the success. However, the correlation coefficient is rather

Table 2.5 Correlation of degree of changes made after launch and success of international market introductions

Hypothesis	Element	Correlation coefficient, r (Degree of changes made after launch and success)	Hypothesis confirmed (✓) rejected (x)
H5a	Definition of target group	n.s.	x
H5b	Definition of positioning	n.s	x
H5c	Definition of product strategy	n.s.	x
H5d	Definition of pricing strategy	−0.256*	✓
H5e	Definition of communication strategy	n.s.	x
H5f	Definition of distribution strategy	n.s.	x

* Significance at the 0.05 level.
** Significance at the 0.01 level.

Table 2.6 Correlation of quality of elements and degree of changes made after launch

Hypothesis	Element	Correlation coefficient, r (Quality of elements and degree of changes made after launch)	Hypothesis confirmed (✓) rejected (x)
H6a	Definition of target group	n.s.	x
H6b	Definition of positioning	−0.305*	✓
H6c	Definition of product strategy	−0.297*	✓
H6d	Definition of pricing strategy	−0.422**	✓
H6e	Definition of communication strategy	−0.397**	✓
H6f	Definition of distribution strategy	n.s.	x

* Significance at the 0.05 level.
** Significance at the 0.01 level.

small, indicating a weak influence of pricing strategy. Thus, Hypothesis 5 is rejected.

Additionally, we wanted to look at the impact of the quality of elements of international market introduction and changes made after the launch proposing a positive impact in Hypothesis 6. Results confirmed that with the exception of target group definition and distribution strategy, there is a significant positive correlation between the quality of elements and the degree of changes made (Table 2.6). Therefore it can be confirmed for the majority of elements: the higher the quality, the lower is the degree of changes that have to be made after launch. As explained earlier, target group definition and distribution strategy are basically fixed, thus they won't be changed after the launch in general even if the product is doing very badly. Thus Hypothesis 6 can partly be confirmed.

Assessment of additional success factors

In order to explain the degree of influence of cooperation on the success of international market introductions we wanted to look more closely at two additional factors of possible influence: participation in decision-making and cultural-geographical distance.

Participation of employees has been identified as a critical success factor in implementing global marketing strategies (Belz, Müller, and Senn, 1999). Therefore we assume that cooperation itself might not be sufficient. Probably it is equally important that local product management can actually participate in the decision-making process and can truly influence the diverse tasks of an international market introduction.

Table 2.7   Influence of participation on success of international market introductions (n=65)

Decision for... were made by	Success of international market introduction (Mean from 1–5)		
	Central unit only	Both together	Local product management only
Definition of target group*	3.1	3.7	3.9
Definition of positioning*	3.1	3.8	4.0
Definition of product strategy*	3.1	3.7	4.1
Definition of pricing strategy*	2.9	3.5	3.8
Definition of communication strategy*	2.7	3.6	3.7
Definition of distribution strategy	3.0	3.1	3.6

One-Way ANOVA, F-Test, *Significance p< 0.05.

Thus we divided the cases into three groups: in the first group every decision was made by the central unit only, in the second group both central
unit and local product management were deciding together, and in the third
group decisions were made solely by local product management. We then
calculated mean scores for the success of the international market introduction for each group individually and probed for significant differences
applying a F-test (Table 2.7).

The analysis shows that the mean scores of the success are significantly
different between the groups apart from the distribution strategy. Again,
the distribution strategy is basically fixed and therefore no joint decision-
making is required, but for all other elements it can be confirmed that
participation in decision-making significantly influences success. When
looking at the mean scores for success it becomes evident, that they are
lowest in the group where the decisions were made by the central unit only.
Success of international market introduction was higher when both units
together or the local product management only were deciding. Even though
we were asking only local product managers from the German speaking
area, we were evaluating international market introductions that could have
been initiated and managed in every country of the world. Looking at the
location of the central unit that was responsible for initiation and management, almost one third of the evaluated international market introductions
was managed from Germany. In approximately 55 percent of the cases the
location of the central unit was in one or more of the European countries,
including France, the UK, Belgium, Netherlands, Italy, Denmark, and Spain.
In about 15 percent of the cases the international market introduction was
managed from overseas, for example from the the US, Canada, and Japan.
Quite often cooperation is influenced by the cultural differences and geographical distances of the parties involved. Since Austria and Germany share
a common language and cultural background, it is probably easier to manage a market introduction for Austria centrally from Germany rather than
from Japan. Thus the cultural-geographical distance between the central

Table 2.8 Influence of cultural-geographical distance on success of international
market introductions (n=63)

**International market introduction was	Location of central unit	
	In Europe	**Outside Europe**
A high/very high success	61%	30%
A failure/total failure	7%	70%
Neither a success nor a failure	32%	0%

One-Way ANOVA, F-Test, **Significance $p < 0.01$.

unit and the local product management might as well have an influence on the success of international market introductions.

To prove this assumption we divided the sample based on the location of the central unit in one group where the market introduction was managed from Europe and another group where it was managed from outside Europe. Using a F-test we looked for significant difference between these groups with regard to success. The results were positive (Table 2.8). The success rate was significantly higher for the international market introductions that were initiated and managed by a central unit located in Europe.

Therefore we can conclude that cultural-geographical distance between the central unit and local product management significantly influences the success of international market introductions.

Summary and conclusion

MNCs generally try to initiate and manage international market introductions centrally to use synergies and save cost. Yet this central and standardized approach requires a sharing of roles and responsibilities between a central unit responsible for initiation and management of the international market introduction and local product management responsible for the actual launch of the new product in its local market. But this approach bears the risk that local market needs are not taken into account sufficiently, coordination problems arise, and rejection amongst local managers appears. Quite often, this leads to an ultimate failure of the new product.

Based on these assumptions this study examined the impact of cooperation with local product management on the success of international market introductions initiated and managed centrally.

Survey results show that the simple act of cooperation itself is not influencing success. But if cooperation takes place, then the quality of cooperation is critical for the success of the international market introduction. The quality of cooperation is impacting upon success directly but also indirectly. The higher the quality of cooperation, the higher the quality of the elements of international market introduction are and this quality of elements has a positive impact on success. Both the direct and the indirect influence were confirmed for almost all tasks of the process, but especially for the ones in the early phases. Therefore, the central unit should cooperate with local product management right from the beginning. In addition, participation of local product management in decision-making has also been identified as a success factor of international market introductions. Thus, cooperation with local product management alone is insufficient – local product management has to be able to take part in the decision-making process, too. The success of international market introduction was also influenced by the cultural-geographical distance between the central unit and local product management: the higher the cultural-geographical distance the higher the failure rate, probably because local market needs were not taken into account sufficiently. The closer the central unit, the more likely they can obtain local knowledge, but the higher the distance the more important is it to use the local product management as a profound source of detailed market knowledge.

Thus based on the research results we can give the following recommendations for successful management of international market introduction:

The higher the cultural-geographical distance of the central unit, the more it should cooperate with local product management;

When the central unit cooperates with local product management, the cooperation needs to be of a high quality and take place without exception for all tasks of the process;

Local product management should participate in all decisions with regards to all elements of the international market introduction.

While this study has shed light on the influence of cooperation on success in international market introduction of new products, some potential limitations exist. First of all the results are limited to the consumer goods industry in the German speaking area. Further research might want to expand the framework to other industries and include local subsidiaries across a range of different countries. With regard to the sample we have to consider that the approach to boost sample size by sending three questionnaires to each subsidiary bears the risk of distorting the sample, because subsidiaries who send back three questionnaires are slightly overvalued compared to those who send back only the one. Even though this research has clearly identified cooperation with local product management as a success factor of international market introductions we have to point out that higher levels of cooperation lead to higher levels of coordination required, for example more travel, more phone calls, more meetings, which could impact success negatively. Probably an optimum exists and further research should be looking into this. As a last word of caution we have to highlight that successful management of international market introductions may not only focus on cooperation but has to consider all other success factors as well.

Managerial implications

- Globally active firms typically initiate and manage international product market introductions centrally in order to use synergies and save costs. Such a central and often standardized approach requires a well planned and coordinated sharing of roles and responsibilities between the central unit and the local product management in order to fully understand local market needs and requirements and to avoid acceptance problems or negative attitudes of local sales staff.
- Our research shows that cooperation as such among central and decentral, local product management is often insufficient and does not automatically

lead to success. A successful market introduction is rather dependant upon how cooperation is being managed (quality of communication, management style, and timing, etc.).

- The higher the quality of these various elements and facets between central and decentral units, the higher the likeliness of success of the market introduction.
- Therefore the central unit should closely cooperate with the local product management right from the beginning of market introduction project.
- Further close participation of local product management in the decision-making process has a strong impact on the later success of an international market introduction.
- Cultural distance is another important factor of influence: The higher the cultural distance between the central and local units the more important it becomes to use local knowledge as a source of information.

References

Belz, C., Müller, M., and Senn (1999). *Die Implementierung globaler Marketing-Strategien in Industriegüterunternehmen: Ergebnisse einer explorativen Untersuchung.* St. Gallen: Thexis.

Benedetto, A. C. (1999). "Identifying the Key Success Factors in New Product Launch." *Journal of Product Innovation Management* 16(6): 530–44.

Boutellier, R., Corsten, D., and Lach, C. (1997). "Neue Ansätze im Projektmanagement der Produkteinführung." *Thexis* 15(3): 13–21.

Boutellier, R. and Lach, C. (2000). *Produkteinführung: Herausforderung für Marketing und Logistik.* München: Hanser.

Call, G. (1997). *Entstehung und Markteinführung von Produktneuheiten: Entwicklung eines prozeßintegrierten Konzepts.* Wiesbaden: Gabler.

Chryssochoidis, G. and Wong, V. (1998). "Rolling Out New Products Across Country Markets: An Empirical Study of Causes of Delays." *Journal of Product Innovation Management* 15(1): 16–41.

Goshal, S. and Bartlett, C. A. (1988). "Creation, Adoption, and Diffusion of Innovations by Subsidiaries of Multinational Corporations." *Journal of International Business Studies* 19(3): 365–88.

Haake, S., Moore, C., and Oliver, N. (2000). "Product Innovation in a Mature and Domestically Oriented Industry: The UK and German Food Manufacturing Industry." 7th International Product Development Management Conference (May 29–30).

Kahn, K. B. (2001). "Market Orientation, Interdepartmental Integration, and Product Development Performance." *Journal of Product Innovation Management* 18(5): 314–23.

Lach, C. K. (2001). *Gestaltungsempfehlungen zur Einführung komplexer Neuprodukte.* Aachen: Shaker.

Langerak, F., Peelen, E., and Commandeur, H. (1997). "Organizing for Effective New Product Development." *Industrial Marketing Management* 26(3): 281–9.

Lawrence, P. R. and Lorsch, J. W. (1967). *Organization and Environment: Managing Integration and Differentiation.* Boston, MA: Division of Research. Graduate School of Business Administration. Harvard University.

Meffert, H. and Bolz, J. (1998). *Internationales Marketing-Management.* Stuttgart: Kohlhammer.

Murphy, C. (2003). "Innovation Masterminds: How Do fmcg Giants Organise Their New Product Development?." *Marketing* (May 15, 2003): 24–25.

Pinto, M. B. and Pinto, J. K. (1990). "Project Team Communication and Cross Functional Cooperation in New Program Development." *Journal of Product Innovation Management* 7(3): 200–12.

Rothwell, R. (1985). "Project Sappho: A Comparative Study of Success and Failure in Industrial Innovations." *Information Age* 7(4): 215–9.

Rothwell, R., Freeman, C., Horsley, A., Jervis, V., Roberson, A., and Townsend, J. (1974). "Sappho Updated: Project SAPPHO Phase II." *Research Policy* 3(3): 258–91.

Sauter, M. (2001). *Internationale Markteinführung technologischer Innovationen: Eine Management-Konzeption.* Aachen: Shaker.

Schreyögg, G. (1999). *Organisation: Grundlagen moderner Organisationsgestaltung.* Wiesbaden: Gabler.

Schröder, H. (1994). "Erfolgsfaktorenforschung im Handel: Stand der Forschung und kritische Würdigung der Ergebnisse." *Marketing ZFP* 16(2): 89–105.

Unilever (2002). Geschäftsbericht 2002, http://www.unilever.de/40/UnileverGB_ 2002.pdf; accessed January 17, 2005.

Zentes, J. (1992). "Organisation der Handelsbetriebe." In E. Frese, E. (Ed.), *Handwörterbuch der Organisation.* Stuttgart: Poeschel: 755–70.

Appendix

	Cooperation between the central unit and local product management in terms of...	Analysis of consumer needs	...can positively influence the success of international market introductions.
H1a		Analysis of consumer needs	
H1b		Analysis of competition	
H1c		Definition of objectives	
H1d		Definition of target group	
H1e		Definition of positioning	
H1f		Definition of launch date	
H1g		Definition of product concept	
H1h		Definition of brand name	
H1i		Design of packaging	
H1j		Definition of product range	
H1k		Definition of pricing strategy	
H1l		Definition of communication strategy	
H1m		Definition of distribution strategy	
H1n		Hand over to sales force	
H1o		Definition of launch budget	
H1p		Measurement of development	

	Cooperation between the central unit and local product management in terms of...		...can positively influence the quality of the...	
H2a		Definition of target group		Target group
H2b		Definition of positioning		Positioning
H2c		Definition of product strategy		Product strategy
H2d		Definition of pricing strategy		Pricing strategy
H2e		Definition of communication strategy		Communication strategy
H2f		Definition of distribution strategy		Distribution strategy

	The higher the quality of...	Definition of target group	...the higher is the success of international market introductions.
H3a		Definition of target group	
H3b		Definition of positioning	
H3c		Definition of product strategy	
H3d		Definition of pricing strategy	
H3e		Definition of communication strategy	
H3f		Definition of distribution strategy	

	Cooperation between the central unit and local product management in terms of...		...can lower the degree of changes made to the...after launch.	
H4a		Definition of target group		Target group
H4b		Definition of positioning		Positioning
H4c		Definition of product strategy		Product strategy
H4d		Definition of pricing strategy		Pricing strategy
H4e		Definition of communication strategy		Communication strategy
H4f		Definition of distribution strategy		Distribution strategy

Continued

H5a	The fewer	Definition of target group	... the higher is the success	
H5b	changes are	Definition of positioning	of international market	
H5c	made to ... after	Definition of product	introductions.	
	launch.	strategy		
H5d		Definition of pricing strategy		
H5e		Definition of communication		
		strategy		
H5f		Definition of distribution		
		strategy		
H6a	The higher the	Definition of target group	... the fewer	Target group
H6b	quality of ...	Definition of positioning	changes	Positioning
H6c		Definition of product	are made to	Product strategy
		strategy	the ... after	
H6d		Definition of pricing strategy	launch.	Pricing strategy
H6e		Definition of communication		Communication
		strategy		strategy
H6f		Definition of distribution		Distribution
		strategy		strategy

3

M&A and Innovation: The Role of Relatedness between Target and Acquirer

Bruno Cassiman, Massimo G. Colombo, and Larissa Rabbiosi

Introduction

The number and value of mergers and acquisitions[1] (M&As) have continued to increase over the last years, both in relative and absolute terms, mainly due to a series of technological, economic, and political changes ranging from the diffusion of Information and Communication Technologies (ICTs) to globalization of markets, liberalization, and privatization processes. In addition, as global competition intensifies, innovation has become a more crucial source of strategic competitive advantage. Within this context, over the last decades there has been an upsurge of interest among scholars on the importance of understanding the consequences of M&As on companies' innovation strategies and performance (Ahuja and Katila, 2001; Capron, 1999; Hagedoorn and Duysters, 2000; Hall, 1990; Hitt et al., 1991). The results are often conflicting and fail to find any robust evidence (for an exhaustive survey, see Veugelers, 2006). These mixed findings are mainly due to the fact that the total effect of an M&A on different measures of innovation – inputs, outputs, and performance – can increase or decrease depending on the forces that dominate the M&A. Therefore, it is very important to identify specific characteristics of an M&A in order to analyze its relationship with the post-deal innovation activity of the merging firm. Furthermore, while initial conditions matter for the innovation outcome of the M&A, we also argue that the management response during the post-acquisition integration will affect the final outcome of the M&A.

In this chapter, we focus on horizontal M&As (that is M&As between firms that operate in the same industries). Following the work of Cassiman and Colombo (2006) we argue that among the above characteristics, relatedness between the acquiring and target firms figures prominently. In particular, following Cassiman et al. (2004) we contend that the impact of an M&A on the innovative activity of the new merged entity depends on both technological relatedness and market relatedness. The former can be defined as the degree of similarity between the merging firms in technological resources, while the latter is the degree of similarity between the merging firms in product markets, geographic markets, and customer base. In addition, social relatedness between the two firms, measured by the existence of previous collaborative relations between them (that is supplier-customer relations or other technology or non-technology collaborative relations) is likely to influence the measurable impact of the M&A on the innovative activities of the merged firm.

Within the next pages we classify M&As according to their market and technological relatedness. Specifically, we identify three types of M&A coming out from the interaction of both technological and market relatedness and we compare the M&A's impact on the innovation process across the different M&A types. Then we analyze how premerger collaborative relationships between the target and acquiring companies might moderate this impact. Finally, we present empirical evidence based on three in-depth case studies that fit different M&A types and show how the expected innovation outcome based on the initial relatedness conditions is enhanced by an appropriate response by management during the post-acquisition integration.

Direct and indirect impact of M&As on innovation: the role of market and technological relatedness

As far as market relatedness is concerned, a distinction must be made according to whether the acquiring and target firms are rivals or non-rivals. Acquirer and target firms are considered as rivals when before the merger they had the same product lines and operated in the same geographical markets.

As mentioned before, M&As can also be categorized according to technological relatedness. Specifically, in several horizontal M&As merging firms were active in overlapping R&D projects before the deal; these M&As are assigned to the "same technology" category. In contrast with this case there will be M&As in complementary technology fields: before the deal acquirer and target firms were active in different technology fields but competences and knowledge that they had developed can be transferred and combined in the merging firm so as to obtain synergistic gains. Similar gains can be obtained in M&As with merging firms that used to operate in the same technological fields, but in different stages of the R&D process; accordingly they are included in the "complementary technology" category.[2]

Following Cassiman and Colombo (2006), combining the above two classifications it is possible to define four M&A types, as indicated in Figure 3.1. Quadrant I represents M&As that combine rival firms with overlapping technology portfolio, while in Quadrant II there are M&As between acquirer and target firms that were non-rival but had similar technological activities.

Quadrant III captures merging firms that were non-rival and active in complementary technologies, while M&As between rival firms with complementary technology portfolios emerge in Quadrant IV. As the last scenario is quite unlikely to occur, we consider the first three M&A types to investigate the effects of M&As on innovation.

We start examining how technological relatedness may explain the effect of an M&A on the innovative activities of the new merging firm. If the target and acquiring firms had similar technological operations (column 1, Figure 3.1) we would expect these operations to exhibit increasing returns to scale, when they are combined. Moreover R&D rationalization through the reduction or elimination of common inputs (that is R&D personnel or R&D laboratories) for the production of innovative output, is likely to occur. The possibility of exploiting similar competences could increase the scale of the R&D projects, and R&D tasks between target and acquirer may become more specialized with this leading to a productivity increase in the R&D function.

However, it should be recognized that some diseconomies in the R&D process might surface as the organization grows larger. In particular, it may be difficult in large and bureaucratic R&D organizations to stimulate the creativity of the R&D personnel. Moreover, R&D rationalization may be a source of organizational turmoil in R&D laboratories, with the effort and attention of scientists and technologists being diverted from productive technological activities.

Considering the complementary technology category (column 2, Figure 3.1), we would expect the acquirer and the target companies to be able to share their competences and knowledge across different R&D projects. In other words, one is likely to observe resource redeployment leading to the creation of knowledge and capabilities that did not exist before. M&As between firms with complementary technological specialization might be a vehicle for the technological diversification of the merging firm. Therefore, the latter could exploit economies of scope, increase R&D outputs, and improve R&D performance.

Let us now turn the attention to market relatedness. In principle, an horizontal M&A may engender an increase in output in so far as it significantly increases the efficiency of the combined firm. In turn, the achievement of economies of scale and/or scope in production and/or distribution activities will lead to increased post-merger R&D efficiency, as R&D fixed costs are spread over more/different output. This effect applies to both rival and non-rival firms. Nevertheless, a merger between rival firms will also increase the market power in the output market of the merging firm leading to a reduction in output (and higher prices); whether this creates greater or lower incentives to R&D investments is questionable (see Aghion et al., 2005). However, if the merger creates barriers to entry in technology, we would observe a reduction in competition in the technology market: for instance, the elimination of a competing innovative product standard that might lead to both lower R&D investments levels and lower R&D perform-ance. Accordingly, the a priori predictions are that M&As between rivals are likely to have a more negative (or less positive) effect on the R&D inputs and outputs of the merging firm compared to M&As between non-rivals. In addition, in the latter case we would expect less restructuring activities and more exploration of new technological fields.

In general, market relatedness and technological relatedness are poten-tially correlated.[3] Therefore, in order to highlight the influence of techno-logical relatedness on the R&D process of the merging firm, one should focus attention on M&As between non-rival firms (Quadrants II and III in Figure 3.1). We expect changes in the R&D process of the merging firm to be more positive for firms in the complementary technology category (Quadrant III) than when acquirers and targets are active in the same tech-nologies (Quadrant II). In fact, in the latter case the organizational costs of R&D rationalization may offset the benefits arising from R&D scale econo-mies. Conversely, these costs are absent in the former situation when firms benefit from the synergistic combination of complementary technological capabilities. Similarly, the role of market relatedness can be isolated focus-ing on the subset of firms that before the M&A had R&D operations in the same technologies (Quadrant I vs. Quadrant II). In both situations, targets and acquirers are likely to have overlapping R&D processes. Therefore, when the aggregation of markets induces a larger output, the R&D activities of the merging firm would benefit from economies of scale both through a stronger specialization and elimination of duplication (that is a cut in R&D personnel, termination of concurrent projects, closing of R&D facilities, etc.). Nevertheless, as mentioned before, when firms are rivals, the greater market power of the merging firm may hamper technological competitive-ness, with detrimental effects on the R&D process.

The role of social relatedness

In addition to market and technological relatedness, we need to high-light the importance of an other dimension for understanding the effects of M&As on innovation: social relatedness. We mentioned above that the alleged positive effects of a M&A on the innovation activity of the com-bined entity may be jeopardized by the need for post-deal restructuring and the associated organizational troubles. In this perspective, the pre-M&A collaborative relations between acquirer and target firms play a key role in facilitating the integration of the merging organizations and the exploita-tion of the existing knowledge and competencies due to greater trust and mutual understanding (Porrini, 2004).

In fact, high internal resistance to the organizational changes engendered by the M&A can slow the decision-making process and damage the inno-vation potential of the merging firm. When acquirer and target firms do

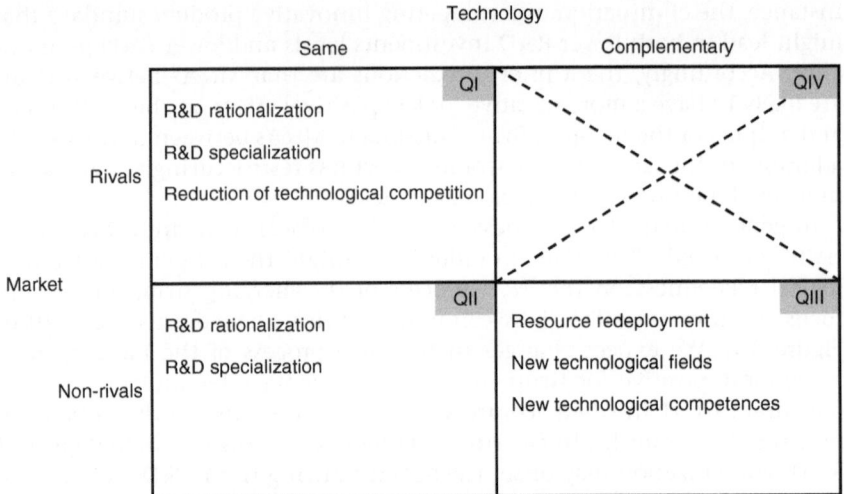

Figure 3.1   M&A classification based on technological and market relatedness
Source: Elaboration from Cassiman and Colombo, 2006.

not know each other before the deal agency problems (Jensen, 1986) are more likely to influence the motivation of researchers at the new entity, in turn negatively affecting R&D inputs and performance. As it was mentioned before, post-merger organizational problems are expected to be more serious for firms with the same rather than complementary technologies, as well as for rivals rather than non-rivals firms. Therefore, prior collaborative relationships would play a prominent role for M&As especially in Quadrant I. In this case, the fact that the partners had the opportunity to get to know each other before the deal and to assess ex-ante the merging firms' competencies and their fit, would facilitate integration and ensure a more likely success of the M&A (Gerpott, 1995).

Empirical evidence

In this section we briefly present three case studies of M&As in the chemical and pharmaceutical industries, based on interviews with firms' high level managers. The aim of the following analysis is to better explore the process that explains the impact of M&As on the innovative activities of the merging firm for different types of M&As.[4] In addition we show how managerial decisions during the post-acquisition process positively affected the innovation outcome of the merger. The three cases cover the Quadrants I–III of Figure 3.1, which is based on technological and market relatedness. Specifically, the Novamax-Henkel and Solvay-BASF deals allow us to investigate M&As between acquirer and target companies that were rivals and non-rivals respectively, and were active in the same technological fields. The DSM and Girst-Brocades case illustrates a deal between non-rival firms that were active in complementary technological fields.[5]

The acquisition of Novamax by Henkel

In 1996 the German multinational company Henkel acquired Novamax, a division of the Canadian Molson Group. Novamax was founded in 1988 and operated in the development of surface treatments for metals. At the time of the deal, the Henkel Group was organized along the following product divisions: Chemical Products, Metal Chemicals, Industrial Adhesives and Technical Consumer Products, Cosmetics and Toiletries, Detergents/Household Cleansers, and Industrial and Institutional Hygiene.

As to market relatedness, Novamax operated in the same metal chemical business as Henkel, but had a different geographical focus. Although Henkel dominated the surface technology industry, some competitors could have threatened this dominance with the acquisition of an important player as Novamax. Therefore, acquiring Novamax permitted Henkel to reduce the probability of having a competitor much closer in size in this market. The two firms operated with a common production technology and accordingly we can place them in the same technology category.

As suggested by the arguments raised in the previous sections, market relatedness created room for rationalizations in administration, sales, and manufacturing. Specifically, three out of five Novamax manufacturing plants were closed and the Novamax administration was shut down. Of the top management team of Novamax, 50 percent was retained, 30 percent left voluntarily, and the remainder was forced to leave Henkel after the acquisition. As a result of the technology overlapping, only Novamax R&D projects complementary to those of Henkel were transferred to the Henkel R&D labs. One of the two Novamax R&D labs was closed, about 10 percent of R&D personnel was cut, and the remaining researchers were transferred to other locations. About 2000 out of 3500 products resulting from the combined portfolios of the acquirer and the target firm, were retained.

This rather radical rationalization may have created organizational troubles, hampering the productivity of R&D. However, through the achievement of critical mass in technological fields already covered by the two companies, R&D became more focussed on specific technological fields and scientists more productive. A crucial aspect of the successful post-acquisition restructuring was also the recognition by Henkel of the know-how of Novamax and its well-organized R&D approach. For instance, Novamax was fast and efficient in rolling out new products. After the acquisition, Henkel's Metal Chemicals Division attempted to assimilate the above capabilities, and this integration effort played a prominent role in enhancing motivation within the R&D personnel of the merging firm.

Finally, the combinations of markets allowed to improve the product line, a renewed focus on the client played an impressive role in attracting new clients, the increase of strengths in marketing and sales generated efficiencies. The merging firm was able to exploit potential economies of scale and scope, and finally technological performance was not hurt.

The acquisition of BASF's vinyl activities by Solvay

Solvin started to operate in 1999 as a joint venture between Solvay (75 percent) and BASF (25 percent). Specifically, Solvay acquired the vinyl activities of BASF and the two companies merged their competencies in polyvinyl chloride (PVC) and polyvinylidene chloride (PVDC) under Solvin. Solvay is an international chemical and pharmaceutical group, headquartered in Brussels (Belgium). The vinyl products represented approximately 45 percent of the sales of the Plastics Sector before the merger. BASF was founded in 1865 and at the time of the deal, its activities were divided into five segments: chemicals, plastics and fibers, colorants and finishing products, health and nutrition, oil and gas. The vinyl activities resorted under the plastics and fibers operating division.

Solvay and BASF were active in the same business and in overlapping geographical markets with a 15 percent and 5 percent market share, respectively. Nevertheless, they were not direct competitors: BASF focused more on larger firms, whereas Solvay targeted small and medium processing companies. The product mix was also not really overlapping in the sense that Solvay was more specialized in commodities and BASF more in specialties.

In terms of technology relatedness, both companies were operating in the same technological field. As a consequence, the merging firm exhibited the expected increasing returns to scale, as well as a consequent R&D rationalization through the reduction or elimination of common inputs. Specifically, with the realization of Solvin, one BASF plant was closed, more than hundred workers were dismissed, and both companies experienced a cut in R&D personnel. For Solvay this cut amounted to approximately 20 percent, for BASF it was larger (about 30 percent). Two BASF teams were transferred from BASF to Solvin. Mainly on the BASF side nonconcurrent R&D programs were terminated, while there were no concurrent R&D programs. After the deal, the research consisted of one centrally managed team in Brussels.

Although a rather drastic rationalization process took place in both merging firms, Solvin increased its technological performance and enhanced the efficiency of R&D operations. Since before the deal the vinyl business was a tiny fraction of the activities of both BASF and Solvay, the creation of Solvin radically increased the survival chances of the vinyl business for each of the two companies. Accordingly, the focus on the new core business boosted the motivation of the R&D personnel working on vinyls, and their technological performance improved impressively. Finally, it is worth noting that with the establishment of Solvin, it was also possible for both BASF and Solvay to increase their total market share in this business.

The acquisition of Gist-Brocades by DSM

In 1998 DSM acquired Gist-Brocades, a group that at the time of the acquisition developed and manufactured intermediate products for the pharmaceutical and food industries. In particular, Gist-Brocades was the world's number one supplier of antibiotics. DSM was founded in 1902 as the Dutch national mining company. However, this company has changed so much over time, in the attempt to respond to the challenges and opportunities of the economic and industrial world. As a result of this process, DMS developed know-how in the fields of value-added processes and products, particularly products for the pharmaceutical and the food industries, and of high performance materials. At the time of the deal, DSM's activities were grouped into three clusters: Life Science Products, Performance Materials and Polymers, and Industrial Chemicals.

Since the two companies operated in antibiotics or intermediates necessary for the production of antibiotics with a different product mix, no horizontal overlap in product markets existed before the acquisition. Moreover, they served the same clients but with different products. Accordingly, the two firms can be considered as non-rivals. For what concerns technological relatedness, Gist-Brocades had strong positions in biotechnology – for example in enzymatic routes and fermentation processes – and DSM in chemical processes. Therefore, the two firms were active in complementary fields.

It should be noted that the two firms had previously entered into a 50/50 joint venture (Chemferm). The venture provided mutual access to the research capabilities of Gist-Brocades and DSM on cephalosporins and other pharmaceutical products, and allowed the two firms to deeply know each other before the deal. This premerger collaboration enhanced the mutual learning and the integration of the merging firms after the acquisition. The prior experience in Chemferm facilitated the organization and management of R&D through the combination of the best practices that existed before within the two companies and the use of an ad hoc integration group.

Overall, it was a successful deal. The aggregation of the two non-rival markets – DSM became a leading company in antibiotics and also entered into new activities such as food specialties – had a positive effects on DSM: 18 percent of sales growth and 13 percent of profit growth were attributed by DSM's top managers to the acquisition of Gist-Brocades. All corporate activities of Gist-Brocades were incorporated in DSM and synergies resulted in total cost savings of approximately €45M. The fact that the two companies were not direct competitors might explain why there was no need to rationalize production activities. The impact of the acquisition on the structure of the R&D activities of the merging firm were limited. No R&D laboratories were closed nor was there a cut of R&D personnel. The complementarity in technological fields played an important role for the achievement of the critical mass in enzymatic and fermentation processes. The exploitation of synergies from the combination of knowledge and capabilities of Gist-Brocades and DSM resulted in the development of new technological competencies and the launch of new R&D programs. In both companies the scope of R&D was broadened.

Conclusions

The aim of this chapter was to highlight how managers of large firms can improve the technological strengths and innovative performance through the acquisition of companies that operate in the same business (i.e. horizontal M&As). In particular, we investigated the channels through which M&As have impact on innovation. As suggested by the management literature, we developed our analysis based on the idea that the effects of M&A on innovative activities are contingent on several moderating factors. Specifically, we focused on different types of relatedness – market, technological, and social – in the pre-M&A situation of the merging companies to understand the innovation potential of the new (post-M&A) entity. The case study design allowed us to uncover how M&A market, technological and social relatedness are linked to innovative inputs, outputs, and performance. In particular, the three selected case studies offered detailed information to characterize the specific dynamic mechanisms linking the effects of M&A on innovation for the following typology of M&As: (1) rival acquiring and target firms operating in the same technological fields, (2) non-rival acquiring and target firms operating in the same technological fields, (3) non-rival acquiring and target firms operating in complementary technological fields.

We found that M&As between firms that have both complementary technological capabilities and a non overlapping product mix, geographic focus or customer base, are likely to be especially beneficial. In fact, the merging firm can take advantage of the synergistic gains that arise from the combination of the complementary capabilities possessed by the target and acquiring company before the acquisition, while avoiding the organizational costs inherent in R&D rationalization. Conversely, when firms have overlapping technological portfolios the benefits of greater scale and specialization of R&D operations must be traded-off against the costs of the reorganization of R&D. In fact, organizational turmoil and voluntary abandonment of key researchers might reduce the beneficial effects of economies of scale and/ or scope. It follows that under these circumstances an M&A may well have negative effects on R&D, with this situation reducing the R&D performance of the combined firm below managers' expectations. This effect is exacerbated when the acquirer and target firms are rivals before the deal, that is, they compete in the same product and geographic markets. In this situation, the deal may create new barriers to entry and reduce technological competition.

This study suffers from a number of limitations. Although we used a rich set of indicators and information to capture changes in the R&D inputs, outputs, performance, and organization in both the acquiring and acquired firms, the fact that we considered only three cases has its own shortcomings. The empirical evidence refers to specific industries (i.e. the chemical and pharmaceutical industries) and the analyzed M&A deals were selected by the firms' managers according to general criteria, among which the most important one was that the M&A had substantial effect on R&D. We considered only horizontal M&As and we did not include cases of hostile takeover. In spite of these limitations, we believe that the study offers new insights on the relationship between M&A and innovation. Accordingly, future works that analyze large and representative samples of M&As should be strongly encouraged in order to confirm our results in a multivariate analysis.

Managerial implications

- This study showed that M&As with different characteristics have different effects on the innovative activities of the post-M&A entity. In particular, managers should take into account that the impact of an M&A on the R&D activities of the merging firm is greatly affected by both appropriate ex-ante selection of the target firm and careful management of the post-M&A integration. The ex-ante selection will be important in driving the potential synergies that can be realized within the M&A, and it is determined by the "relatedness" of the merging firms. This relatedness may have several dimensions: technological; market; social.
- The Solvin and the Hankel-Novamax cases indicated that extensive rationalization of R&D activities occurs when the pre-M&A firms are technological related that is they operate in similar technological fields. The rationalization process can be a source of organizational turmoil; it can reduce employees' motivation and increase the lost of key researchers due to voluntary abandonment. Therefore, a negative effect on the innovative performance of the merged entity is expected. However, managers can influence these results if the post-integration process is properly managed. Strong motivation of the R&D personnel through the diffusion of a common culture and the perception of the M&A as a positive sum game is crucial to smooth the rationalization process, thus positively affecting R&D performance. Investing in greater communication and coordination is also decisive. This was clearly a key strength of the creation of Solvin.
- Managerial action and careful attention to the integration process is even more crucial when the negative effects of technological relatedness are exacerbated by market relatedness. This is evident in the Hankel-Novamax case: the decision of Henkel to adopt the market-based strategy of Novamax

efficiently motivated the employees and avoided the disruption of important organizational routines. Accordingly, the adoption of superior best practices and routines of the target firm might help the integration process through the creation of a positive environment and mutual adaptation, in spite of extensive rationalization of R&D operations.

- Our analysis also lend some support to the view that organizational problems that reduce the realization of potential synergies may be more easily overcome if merging firms have previously established collaborative relations between each other. In this perspective, the existence of previous collaborative links between the two parties of the M&A appears to play a key role, as it reduces the information asymmetry that makes scouting for the proper target difficult; it also facilitates the identification of potential synergies of competencies and technologies between the two merging firms.

Notes

1. In this work we use the term 'merger' and 'acquisition' as synonymous.
2. In principle, the target and acquiring companies may have unrelated R&D operations before the acquisition. As a matter of fact, this situation is unlikely to occur in horizontal M&As. So it is not considered here.
3. The arguments that follow are based on Cassiman et al. (2005).
4. For further empirical evidence on this issue, see again Cassiman et al. (2005).
5. For a more detailed description of these case studies, see Cassiman et al. (2006).

References

Aghion, P., Bloom, N., Blundell, R., Griffith, R., and Howitt, P. (2005). "Competition and Innovation: And Inverted -U Relationship." *The Quarterly Journals of Economics* (May): 701–28.

Ahuja, G. and Katila, R. (2001). "Technological Acquisitions and the Innovation Performance of Acquiring Firms: A Longitudinal Study." *Strategic Management Journal* 22: 197–220.

Capron, L. (1999). "The Long-Term Performance of Horizontal Acquisitions." *Strategic Management Journal* 20: 987–1018.

Cassiman, B. and Colombo, M. G. (Eds) (2006). *Mergers and Acquisitions: The Innovation Impact.* Cheltenham, UK: Edward Elgar.

Cassiman, B., Colombo, M. G., Garrone, P., and Veugelers, R. (2004). "The Impact of M&a on the R&D Process. an Empirical Analysis of the Role of Technological and Market Relatedness." *Research Policy* 34: 195–220.

Gerpott, T. J. (1995). "Successful Integration of R&D Functions after Acquisitions: An Exploratory Empirical Study." *R&D Management* 25: 161–78.

Hagedoorn, J. and Duysters, G. (2000). *The Effects of M&A on the Technological Performance of Companies in High-Tech Environments.* Working Paper: United Nations University-Maastricht Economic Research Institute of Innovation and Technology.

Hall, B. H. (1990). "The Impact of Corporate Restructuring on Industrial Research and Development." *Brookings Papers on Economic Activity, Microeconomics*: 85–124.

Hitt, M. A., Hoskisson, R. E., Ireland, R. D., and Harrison, J. S. (1991). "Effects of Acquisitions on R&D Inputs and Outputs." *Academy of Management Journal* 34: 693–706.

Jensen, M. C. (1986). "Agency Costs of Free Cash Flow, Corporate Finance and Take-Overs." *American Economic Review* 76: 323–9.

Porrini, P. (2004). "Can a Previous Alliance between an Acquirer and a Target Affect Acquisition Performance?." *Journal of Management* 30: 545–62.

Veugelers, R. (2006). "Literature Review." In B. Cassiman and M. G. Colombo (Eds), *Mergers and Acquisitions: The Innovation Impact*. Cheltenham, UK: Edward Elgar: 37–62.

4
Revisiting the Firm's R&D and Technological Ecosystem – A Case from a Large IT Firm

Thomas Durand

Introduction – What is an Ecosystem?

Any organizational entity with some strategic autonomy, typically a business unit, has built a set of complex and intricate relationships with its environment over the years. These relationships and exchanges nurture the organizational unit, while the unit also feeds its partners in the environment in return. The partners who are active in this set of relationships constitute the Ecosystem in which the entity lives and tries to prosper.

These partners cover a wide range of players. They include the suppliers of materials, components and services, the clients of the offerings of the unit, distribution channels and vendors, complementary players in the value chain, competitors who entered some form of collaboration with the unit, consulting and engineering firms, designers or public research centers, professional associations, professional bodies defining norms and standards, and so on. Some of the partners may belong to the same legal entity as the focal organizational unit or may be external to the overall organization around the unit at hand.

One may actually refine this definition of an Ecosystem by deciding whether all partners, both external and internal to the mother organization to which the unit belongs, should be included in the Ecosystem or whether the concept of Ecosystem only applies to partners outside the "legal contour" of the firm. As firms increasingly operate in networks, organizations increasingly seem to adopt network-like structures, where the contour of the legal entities is no longer the key base for strategy. Instead, organizational units in large companies tend to play their game of serving their mother organization in unexpected ways, leveraging all sorts of connections with competent players regardless of who owns them. As this blurring effect seems to be an important feature of recent developments in business structures, especially in the context of large multinational corporations, we shall adopt a rather extensive definition here: The Ecosystem encompasses all partners of the organizational unit, external and internal to the mother organization. (for a typical ecosystem see Figure 4.1).

Through its daily activities, the unit keeps rebuilding and shaping its Ecosystem. This is most often a tacit process, as the management of the unit is not fully aware of shaping the overall architecture of the Ecosystem. In fact, decisions to work with a partner are usually made on a case by case basis to address a specific need when it arises.

The unit and its Ecosystem coevolve as they both influence and nurture each other while being influenced and nurtured in return. Yet, the unit aims at leveraging its Ecosystem strategically, to serve its own goals.

As Reve (1990) puts it: "Strategy is about optimizing the combination of a strategic core and a set of alliances, aimed at building a competitive advantage." From the perspective of an Ecosystem, strategy may be viewed as managing dynamically the frontier of the unit interfacing with its partners in the environment.

In this sense, there is no ideal Ecosystem. There are Ecosystems that more or less fit to the strategic intent of a unit at a certain point in time. In fact, a unit's Ecosystem emerged over the year to solve issues encountered along

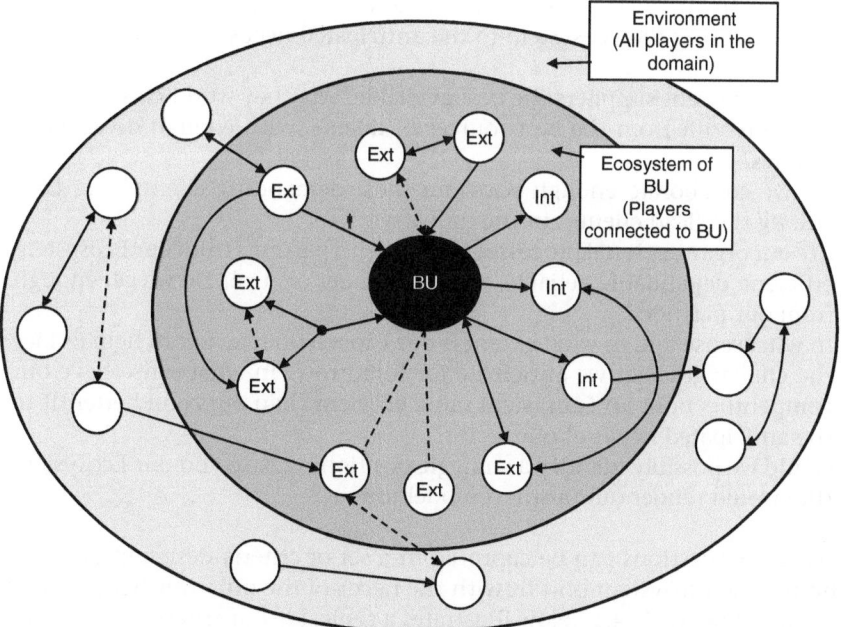

Figure 4.1 A typical Ecosystem of a business unit

the way. Most organizations haven't consciously built their Ecosystems. They have contributed to shape it, though indirectly, through a variety of day-to-day decisions over the years.

New partners were searched for, selected, tried, retained, and maintained through a continuous, most often unplanned process of scanning, negotiation, socialization, and retention. In this sense, the Ecosystem is the result of an emergent process. It was usually not designed and planned in a rational way. It is a legacy of past activities and issues. Current partnerships essentially result from past necessities.

This means that the Ecosystem may not fit the needs of the unit today and, in a changing world, it is unlikely to be fully fit to the issues arising for the future of the unit. Durand, 1992a and b; 1996a and b; 1998; 2001; 2003; 2004a and b; 2006; 2008; Durand and Stymne, 1991; Durand, Mounoud and Ramanantsoa, 1996; Durand and Guerra-Vieira, 1997; Durand Farhi and Brabant, 1997; Durand and Dubreuil, 2001; Gibbert and Durand, 2007; Gonard and Durand, 1994; Kandel, Remy, Stein and Durand; 1991).

Why analyzing the Ecosystem?

As the Ecosystem may not fit the current and future needs of the business unit, it makes a lot of sense to review the set of existing and past relationships that the unit established with its environment. In turn the analysis aims at adapting the Ecosystem to the anticipated needs:

- Are our current suppliers the best available? Are they world class?
- Do we recruit from the best sources of talents, with enough diversity of profiles?
- Do we collaborate enough with our most demanding clients? Are they among the "lead clients" in the industry?
- Are our organizational processes to work with partners from our Ecosystem efficient, dependable, reliable, and fully under control? Do we get enough from our partners?
- In what sense and to what extent is our Ecosystem relevant to help us face the challenges that we anticipate for tomorrow? In what sense have our competitors built an Ecosystem more efficient than ours and better fit to the anticipated needs of our sector?
- Could we possibly identify a strategic positioning based on our Ecosystem that would render our business model unique?

This set of questions can be captured in a set of criteria designed to assess whether the Ecosystem is in fit with the needs of the unit, to what extent it does and how. Table 4.1 below illustrates a typical list of criteria to assess an Ecosystem and build an action plan to better leverage the existing Ecosystem while targeting new potential partners to regenerate the Ecosystem in line with current and future challenges.

Table 4.1 List of criteria to assess Ecosystem

	Criteria to assess the Ecosystem
Input to competitive intelligence	The Ecosystem provides a good access to external sources of information and data
Source of innovation	The Ecosystem provides a good access to sources of ideas (world-class partners, innovative start-ups, etc.)
Influence	The Ecosystem helps influence the environment (e.g. ability to prescribe a specific new technology)
Image	The Ecosystem makes it possible to strengthen the technological image of the unit (events, fairs, symposia, etc.)
Cost optimization	The Ecosystem helps cost cutting (subcontracting, sharing added value steps, etc.)
Short term vs long term	The Ecosystem makes it possible to strike a balance between long term issues (Research partners, scientific scouting) and short term matters (technology transfer, applied R&D)
Geographical coverage	The Ecosystem provides a wide geographical coverage
Quality of internal network	Quality and scope of relations with other internal players in the organization: technology transfer, shared projects...
Quality of external network	Quality and scope of relations with external partners, in line with what the unit needs
IPR	The rules of the game in the Ecosystem ensures adequate IPR for the unit as well as for its partners
Strategic fit	The Ecosystem is in line with the unit's strategy and serves its strategic needs

This is what we exemplify next in the case of the Ecosystem of a R&D unit of a large firm in the business of information technologies (IT).

Analyzing the Ecosystem of an R&D department

This case is about an R&D department in a large IT company with activities in a variety of countries. The R&D unit is part of the R&D labs of the company. It comprises about 100 staff working on a specific topic under the theme of Cryptology in IT.

The case study presented here results from the use of a method requiring seven steps to analyze the Ecosystem of an R&D unit, from the description of the existing Ecosystem to the elaboration of an action plan aiming at better leveraging and regenerating the Ecosystem of this R&D unit.

Step1: List all players which may be regarded as members of the unit's Ecosystem, cluster these into families of partners (subcontractors, clients both internal and external to the mother Group, R&D partners, and so on), assess the intensity of the working relation of each partner to the R&D unit,

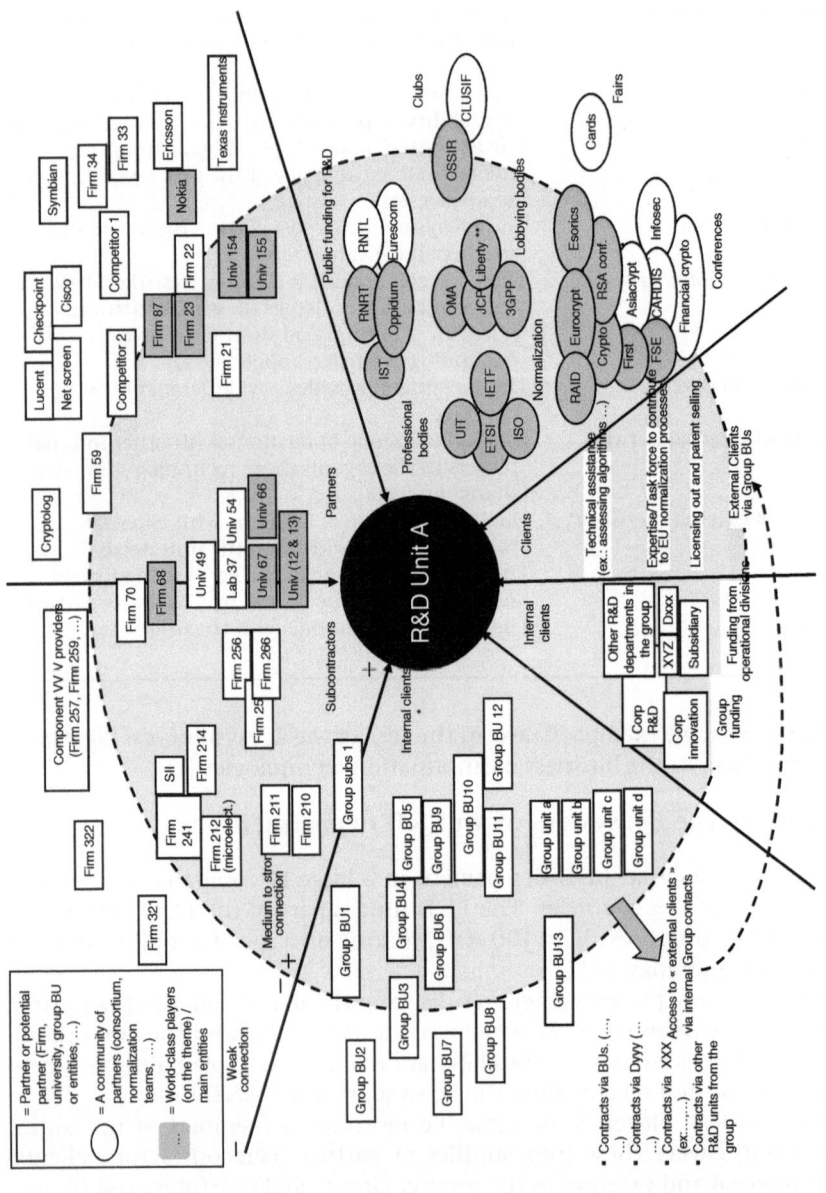

Figure 4.2 Mapping of the Ecosystem

and use this to provide a mapping of the existing Ecosystem. This can be done after a set of interviews with the heads of the R&D unit plus about two or three two-hour group sessions to discuss draft mapping generated along the way, in an iterative process. See the resulting mapping for this R&D unit on Figure 4.2.

Step 2: List the "world-class" organizations which the R&D unit would wish to have (or maintain) in its Ecosystem. Identify explicitly those who are already part of the Ecosystem (and show these on the Ecosystem mapping – see the shaded items on Figure 4.2), those with which preliminary contacts have been established and those with no contact yet. Specify which specific competences these would bring to the unit, or in the case of clients, the extent to which they may pull the unit ahead via lead requirements. Map the external set of world class potential partners beyond the existing Ecosystem, as shown on Figure 4.3 (where the entire Figure 4.2 was shrunk into the inner circle identified as "Ecosystem of the unit"). Numbers on Figure 4.3 indicate world-class players that the unit considers as potentially interesting partners.

Step 3: List the networks of players throughout the field. Identify where the unit is not related to some of the key clubs, clusters, and associations and

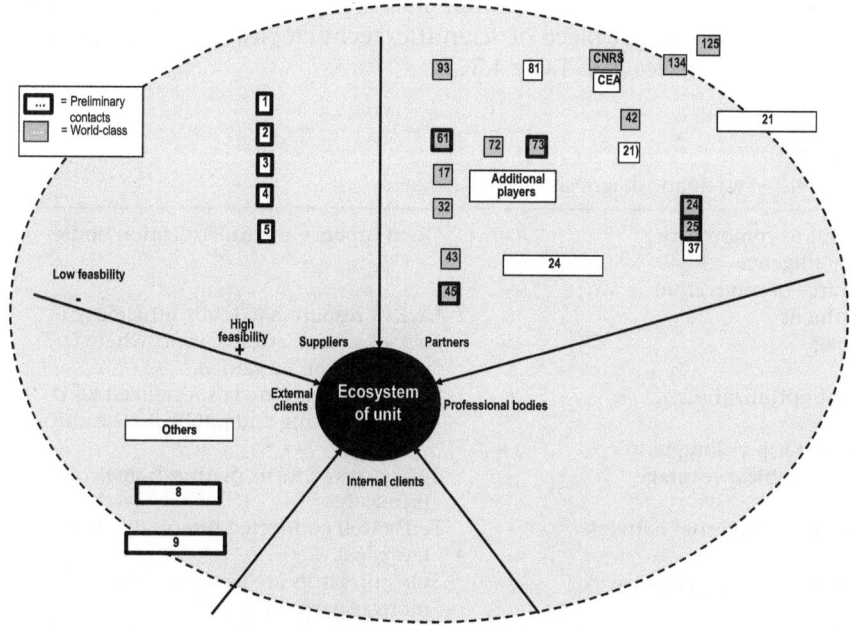

Figure 4.3 Mapping the external set of world-class potential partners

assess when this is an issue. (Part of this is shown under the "Professional bodies" sector of Figure 4.2.)

Step 4: Conduct a similar analysis (Steps 1 to 3) for the major competitors, thus generating mappings similar to those of Figures 4.2 and 4.3. Although not shown here, this was done for three major competitors, based on readily available information complemented by an internet search and some interviews. Although non exhaustive, these provide important insights into the structure of competitors' Ecosystems, their differences, weaknesses, strengths, connections, and overlaps.

Step 5: Formulate a systematic diagnosis of the unit's Ecosystem using the set of criteria shown in Table 4.1. The results of the diagnosis can be recapitulated as shown in Table 4.2.

Step 6: Identify the scientific, technological, and organizational competences (resources, capabilities, and know how) that the R&D unit has. Also identify the scientific, technological, and organizational competences that the unit needs given its objectives and its strategy. (The issue of granularity in such an analysis of competence is not easy to cope with. It is a matter of trial and error to reach the appropriate level of analysis for each specific subtheme of competence) For each piece of competence, assess its stage in the competence life cycle (emerging, pacing, key, basic, obsolete), the relative importance for the unit given its strategy, the level of expertise of the unit on that competence. In addition, identify which players would be the "best-in-class" on that piece of scientific, technological or organizational competence. This leads to Table 4.3.

Table 4.2 Systematic diagnosis of the Ecosystem

Input to competitive intelligence	++	Good inputs via standardization bodies
Source of innovation	++	
Influence	–/+	Lack of resources to lobby efficiently
Image	–/+	Lack of staff to attend events where visibility can be gained
Cost optimization	+	Cost optimization via specialized R&D subcontracting and public R&D funding
Short term vs long term	++	
Geographical coverage	–	Too narrow a focus (within home country primarily)
Quality of internal network	+	Fairly well connected internally (despite a few gaps).
Quality of external network	–/+	Subcontractors are pushing to be treated more as partners
IPR	+	
Strategic fit	?	Strategy needs to be clarified

Table 4.3 Competences of the R&D unit

	Relative importance for unit's strategy	Stage in competence life cycle	Expertise level of the unit	
	L=low, M=medium, H=high	Emerging, pacing, key, basic, obsolete	L=low, M=medium, H=high	Best in class
Competence A	H	Key	HH	Company S12
Competence B	H	Key	HH	Companies B and F.
Competence C	H	Key	H	
Competence D	H	Key	H	Labs XX, YY
Competence E	H	Key	H	
Competence F	H	Key	HH	
Competence G	M	Key	H	
Competence H	H	Key	H	
Competence I	H	Key	H	
Competence J	H	Key	H	
Competence K	H	Key	HH	Firm VVV
Competence L	H	Pacing	M	
Competence M	M	Pacing	M	
Competence N	H	Key	H	
Competence O	H	basic	H	Subcontractor GGG
Competence P	H	basic	H	
Competence Q	L	obsolete	L	
Competence R	M	pacing	H	
Competence S	L	key	M	

In addition, assess which of these competences are too strategic to be developed in partnership with external partners, regardless of them being part of the unit's Ecosystem. Identify those competences where a subcontractor may be invited to bear the competence development and maintenance on behalf of the unit, based on specifications defined and controlled by the unit. Identify those competences where it is acceptable or inevitable to enter an alliance, thus inviting partners to contribute whether already part of the Ecosystem or not. Finally identify the basic and obsolete competences, where there is no longer a potential for any form of leadership or leverage to differentiate and build a technological strategic advantage of any sort as these pieces of competence are widely available to all players in the industry. This leads to Figure 4.4 where competences A1, A2, A3, and so on, are sub-items of competence A shown in Table 4.3.

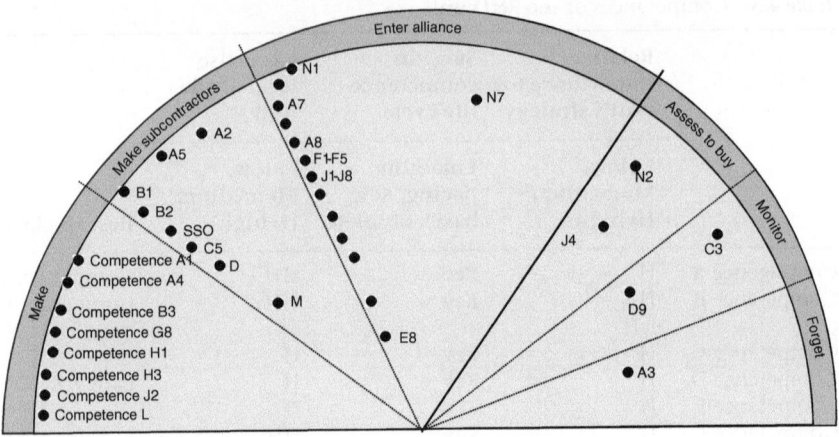

Figure 4.4   Competences map

Step 7: Draw from the analysis above an action plan to better leverage the Ecosystem and to reshape it according to the current and future needs of the unit. The aim is to cover the expected set of competence items where the R&D unit may not be able to build enough competence on its own. On the strategic front, identify new potential partners as targets, with explicit expectations regarding the competence searched for via a potential cooperation. This will help those in charge of approaching these targets. On the organizational front, revisit the process of interacting with members of the Ecosystem to make the best out of cooperating. This is a matter of both work process and attitude in the partnerships. Table 4.4 summarizes the action plan developed in the case of this R&D unit of this large IT Company.

Such an analysis conducted using the method summarized above via the seven steps in fact requires limited time and resources. With the exception of the benchmarking data of Step 4, most of the information needed is readily available in the unit itself. It is thus mainly a matter of collecting the information, via a series of interviews and meeting, plus compiling it into formats that can be directly used to discuss concrete implications for the unit's management. At the end of the seven steps, the various mapping and tables generated provide a useful insight into the Ecosystem of the unit and a concrete action plan to work on transforming the Ecosystem so it's a better fit for the strategic needs of the unit.

Table 4.4 Action plan for R&D

Action / Project	Objectives	Status report
Improve / develop linkages to additional university research centers (, ...)	Mutualize competences and share costs	Some targets have been identified
Enter partnerships with private R&D / industry consortia (...)	Mutualize competences and share costs	
Enter cooperative research agreements with labs from noncompeting Firms	Mutualize competences and share costs	Preliminary contacts with two targets. Promising.
Find partners in sector XXX		Preliminary contacts with Firm T et Lab U
Improve how the Group makes use of the R&D unit's competence	Exploit potential synergies; enhance technology transfer across R&D labs within Group	To be launched
Widen subcontracting scope	Minimize costs	First list of activities to subcontract now available
Enter discussions with lead competitor to offer focused R&D cooperation	Increase our influence via this leading competitor	Contacts ongoing
Gain access to competence A5	Avoid R&D costs and speed up access to competence	

Managerial implications

An Ecosystem tends to appear somehow naturally around the entity as the unit interacts with players in its business environment. Along the way, a variety of ties and linkages are formed which constitute the heart of the Ecosystem. In most cases the Ecosystem was not designed with a clear, organized strategic intent. Instead it resulted from the many interactions with the environment developed over the years as the unit encountered problems and needs that led to find out new external partners.

- There is thus a need to revisit the Ecosystem of any unit in an organized and systematic way to question the set of current partners in light of the unit's current and anticipated strategic needs. It is suggested to conduct such a review of the ecosystem periodically, for example every two years or so, for instance as part of the strategic review of the unit.

- The review is aimed at describing the existing Ecosystem, list the world-class players which the unit would dream to be connected to, compare with the Ecosystem of the main competing units outside, compare the competence that the Ecosystem covers with the anticipated competence needs for the unit, target the potential partners to gain access to, identify ways to improve how the unit leverages its relationship to players from its Ecosystem.

References

Durand, T. (1992). "Dual Technological Trees: Assessing the Intensity and Strategic Significance of Technological Change." *Research Policy* 21(4), 361–80.

Durand, T. (1992). "The Dynamics of Cognitive Technological Maps." In P. Lorange, J. Roos, B. Chakravarty, and A. Van de Ven (Eds), *Implementing Strategic Processes*. Oxford, UK: Blackwell Publishers: 165–89.

Durand, T. (1996a). "Strategizing for Innovation: Competence Analysis in Assessing Strategic Change." In A. Heene and R. Sanchez (Eds), *Competence-Based Strategic Management*. Chichester: Wiley: 127–51.

Durand, T. (1996b). "National Management of Technology and Innovation: Integrating the Firm's Perspective into Government Policies." In H. Thomas and D. O'Neal (Eds), *Strategic Integration*. Chichester: Wiley: SEITEN.

Durand, T. (1998). "The Alchemy of Competence." In G. Hamel, C. K. Prahalad, H. Thomas, and D. O'Neal (Eds), *Strategic Flexibility: Managing in a Turbulent Environment*. Chichester: Wiley: SEITEN.

Durand, T. (2003). "12 Lessons Drawn from 'Key Technologies 2005': The French Technology Foresight Exercise." *Journal of Forecasting* 22(2/3): 161–77.

Durand, T. (2004a). "Selecting Technologies and Themes in Technologies: Clés 2005." In K. Cuhls and M. Jaspers (Eds), *Participatory Priority Setting for Research and Innovation Policy, Concepts, Tools and Implementation in Foresight Processes – Proceedings of the International Workshop on "Futur – The German Research Dialog"*. Stuttgart: Fraunhofer IRB.

Durand, T. (2004b). "The Strategic Management of Technology and Innovation." In D. Probert, O. Grandstrand, A. Nagel, B. Tomlin C. Herstatt, H. Tschirky, and T. Durand (Eds), *Bringing Technology and Innovation into the Boardroom: Strategy, Innovation & Competences for Business Value*. Houndmills: Palgrave Macmillan: 47–75.

Durand, T. (2006). "The Making of a Metaphor: Developing a Theoretical Framework." In J. Löwstedt and T. Sternberg (Eds), *Producing Management Knowledge*. London: Routledge: 179–97.

Durand, T. (2010). "Technology Intelligence." In V. K. Narayanan and Gina O'Connor in C. Cooper, C. Argyris, and B. Starbuck (Eds), *The Blackwell Encyclopaedia for Management 13 – Technology and Innovation Management*, 2nd Edition. New Jersey: Wiley-Blackwell..

Durand, T. and Dubreuil, K. (2001). "Humanizing the Future: Science and Soft Technologies." *Journal of Future Studies, Strategic Thinking and Policies* 3(4): 285–95.

Durand, T., Farhi, F., and Brabant, C. (1997). "Organising for Competitive Intelligence: The Technology and Manufacturing Perspective." In W. Bradford Ashton and R. A. Klavans (Eds), *Keeping Abreast of Science and Technology – Technical Intelligence for Business*. Colombus: Batelle Press: 189–212.

Durand, T. and Guerra-Vieira, S. (1997). "Competence-Based Strategies When Facing Innovation. But What is Competence?." In H. Thomas and D. O'Neal (Eds), *Strategic Discovery: Competing in New Arenas*. Chichester: Wiley: 79–98.

Durand, T., Mounoud, E., and Ramanantsoa, B. (1996). "Uncovering Strategic Assumptions: Understanding Managers' Ability to Build Representations." *European Management Journal* 14(4): 389–98.

Durand, T. and Stymne, B. (1991). "Technology and Strategy in a Hitech Industry." In L. G. Mattson and B. Stymne (Eds), *Corporate and Industry Strategies for Europe. Adaptations to the European Single Market in a Global Industrial Environment*. Amsterdam: Elsevier Science Publishing: 193–215.

Gibbert, M. and Durand, T. (Eds) (2007). *Strategic Networks: Learning to Compete*. Malden, MA: Blackwell.

Gonard, T. and Durand, T. (1994). "Public Research / Industry Relationships: Efficiency Conditions." *International Business Review* 3(4): 469–89.

Kandel, N., Remy, J. P., Stein, C., and Durand, T. (1991). "Who's Who in Technology: Identifying Technological Competence within the Firm." *R&D Management* 21(3): 215–28.

Reve, T. (1990). "The Firm as a Nexus of Internal and External Contracts." In M. Aoki, B. Gustafson and O.E. Williamson (Eds), *The Firm as a Nexus of Treaties*. London: Sage: 133–61.

Part II
Competence

5
Getting Value from Technology: A Process Approach

Clare J. Farrukh, David Probert, and Robert Phaal

Introduction

Technology can represent a major source of competitive advantage and growth for companies. However, integrating technological considerations into business processes effectively is a complex task, requiring the consideration of multiple functions (technical, marketing, finance, and human resources). Technology, combined with highly motivated and properly trained people, enables a business to respond rapidly to changing customer demands and to access and develop new market opportunities.

The challenges associated with the management of technology are compounded by a number of factors, including the increasing cost, complexity and pace of technology advancement, the diversity of technology sources, the globalization of competition and alliances, and the impact of information technology. These challenges also represent a great opportunity for organizations that can fully harness their technological potential.

To compete successfully, companies must assess their technology management strategy and practice, and address how they:

Recognize technological opportunities and threats and convert them into sales and profit.

Exploit existing technology by the effective translation of strategy into operational performance.

Differentiate products using cost-effective technological product and process solutions.

Identify and evaluate alternative and emerging technologies in the light of company policy and strategy and their impact on the business and society.

Reduce the risks inherent in new or unfamiliar technologies.

Harness technology that supports improvement in processes, information, and other systems.

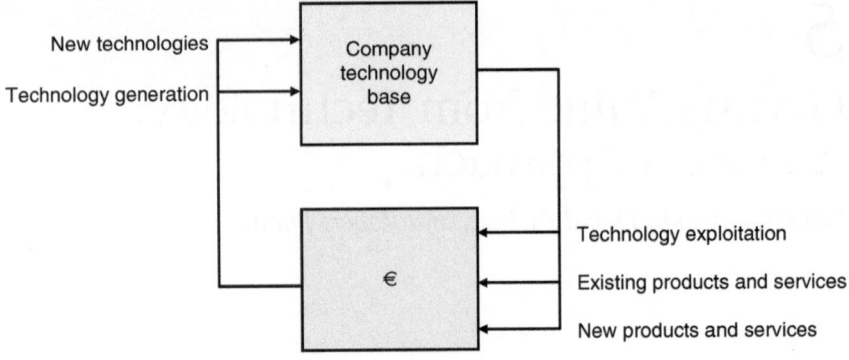

Figure 5.1   Technology generation exploitation cycle

Decrease the "time to market" of new products and services through effective identification and exploitation of technologies that provide competitive advantage.
Protect and exploit intellectual property.

With reference to Figure 5.1, the following five questions must be answered if the full potential of investment in technology is to be realized:

How do we exploit our technology assets?
How do we identify technology that will have a future impact on our business?
How do we select technology for business benefit?
How should we acquire new technology?
How can we protect our technology assets?

These questions form the basis for developing a set of technology management processes in the firm, and are explored further in the following sections, illustrated by case study examples.

How do we exploit our technology assets?

Process perspective

In the competitive marketplace, firms that utilize their technological assets most effectively have a significant advantage. Continued exploitation and renewal of the technology base is essential for long-term survival. The systems that support the delivery of products and services to the market need to be clearly understood, in terms of how technology provides value to the company and its customers – see Figure 5.2.

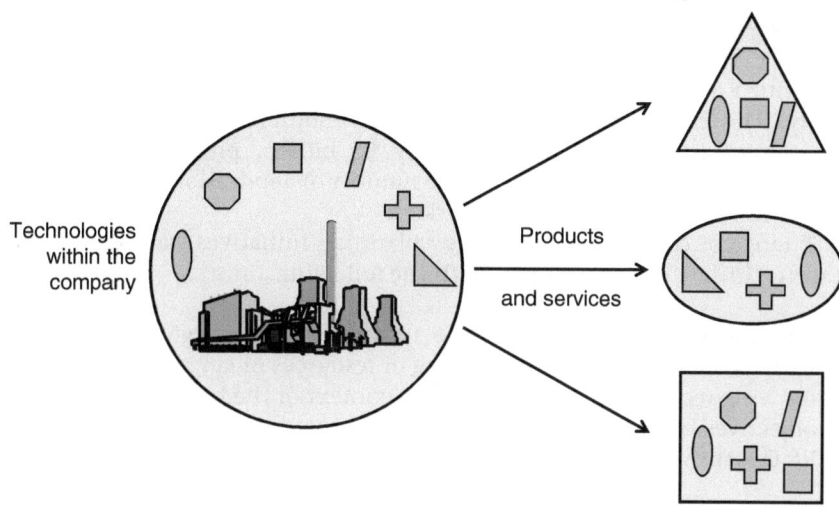

Figure 5.2 Technology exploitation process

Key issues to consider include

Management of the technology base.
Technology planning, including short- to medium-term forecasts of market
  requirements and technological capabilities and trends.
Relationships within the customer-supplier network and with other exter-
  nal sources (for example, standards-making bodies).
Communication channels and information flow.
Operations and resource management.

A clear understanding of the nature of the core technologies in the
company is required. How do these relate to the key skills and capabilities
of staff, products and services, markets and competitor activity?
The company should be aware of the many options available to exploit its
technology, including:

Selling or licensing its technology.
Joint ventures or collaboration.
Technology fusion, whereby existing technologies are combined in innova-
  tive ways to provide new products or services.
Technology transfer processes (internal and external).
Improved business processes and organizational structures to support the
  generation and exploitation of technological capability.

Case study

This example illustrates how cross-business technological synergies were identified and exploited in a large international electronics systems company. The corporation is organized into multiple business units, with an overall turnover of approximately €5 billion, producing high-tech, electronics-based products for a large number of applications in a wide variety of military and commercial markets.

A range of concurrent technology planning initiatives was being undertaken within the organization, with the following aims:

Improving the exploitation of the technological synergies between the various operating units and sharing of resources in key areas.

Improving technology planning in the context of the business / marketing objectives of the firm, and more closely integrating the role of central R&D facilities.

As part of this process, a simple matrix-based method was used to develop a framework to link technological capabilities with business objectives (Figure 5.3). This involved segmentation of the business in terms of technology and business areas in a series of senior management workshops. By ranking and assessing the impact of each technology area on each business area, it was possible to identify core technology areas which are of high value across several business units, and areas of mismatch between value, effort, and risk. This has enabled the organization to focus attention on, and investment in, key areas of common interest and to achieve greater levels of coordination between historically independent business units.

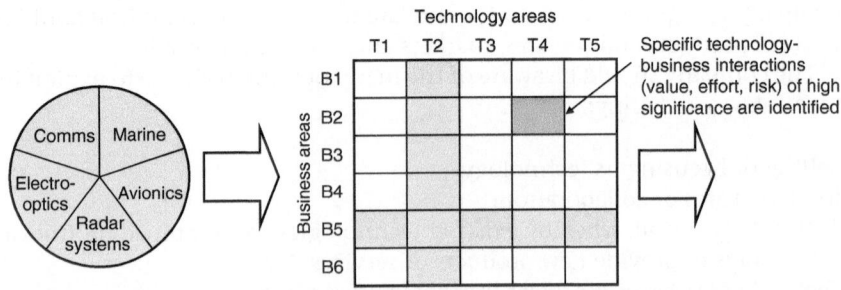

Figure 5.3   Identifying opportunities for exploiting technology synergy

How do we identify technology which will have a future impact on our business?

Process perspective

Maintenance and renewal of the technology base requires that processes be in place for the identification of new technologies that are, or may in the future be, important to the business – see Figure 5.4. This is becoming an increasingly challenging task as the complexity and pace of technological change increase and the sources of technology become more international.

Key issues to consider include

A thorough understanding of the nature and composition of the firm's technology base, in relation to how these capabilities add value to its products and services.

Access to appropriate external and internal networks and sources of information.

Knowledge management systems and communication channels, to ensure that the information is appropriately processed and disseminated.

Technology identification processes include a range of activities:

Systematic scanning of information sources, to develop an awareness of existing and emerging technologies.

Technology and market foresight and forecasting processes, to support the identification and appraisal of emerging technologies.

Monitoring of specific technical areas.

Ways of generating new ideas, to identify new product and process opportunities.

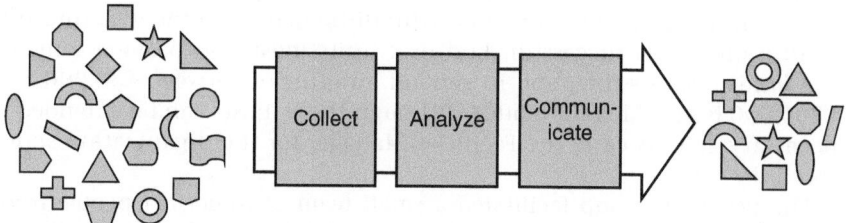

Figure 5.4　Technology identification process

Technical benchmarking to develop an awareness of competitors' capabilities.
Specific data collection in response to new requirements.

In many organizations, technology identification is undertaken on an ad hoc basis, based on informal networks, attendance at trade fairs and conferences, subscription to journals, contacts with suppliers, and so on. In addition to making this more systematic, the challenge is to develop appropriate systems to collate and analyze the data collected, and to disseminate it effectively throughout the organization.

Case study

This example illustrates how a national postal services organization identifies new technology that will impact its operations. The transfer of technology into the business has been assisted by pooling research and expertise into one unit, the Research Group, which explores and identifies opportunities for cost reduction, efficiency improvement, and new products or markets offered by technology.

The Research Group also helps the organization to develop an Innovation Fund – a fast track, specially supported system which uses "ring fenced" funding to launch new ideas and pilot projects.

Most of the Research Group's work is focused on specific areas of identifiable business benefit, but it also consciously scans the world for any technology that may be of use or interest. As part of this research, it came across a revolutionary lighting technology suitable for large open areas, such as mail sorting offices. The technology uses sulphur plasma excited by microwaves to create "white" light, which is then distributed using a "light pipe" that is 20 m long and 25 cm in diameter. To the organization, it offered the potential of flicker-free, daylight spectrum lighting for its staff to work under and reduced power consumption, with obvious benefits to both running costs and the environment. An Innovation Fund proposal was conceived.

The Innovation Fund requires that submissions have business sponsors within the organization who will, ultimately, take the idea into full utilization if it is successful. End user units must also provide some of the funds towards the pilot. In general, funding of between €30,000 and €300,000 is provided for a pilot, although these limits can be extended if necessary. In this case, the Facilities Manager for the organization sponsored the idea.

The Research Group facilitated a small team of interested people from across the company, who took the project from concept to pilot installation in a sorting office. This included visits to existing installations and the loan of a prototype by the chosen supplier.

The key points to take from this study are

Consciously scan outside the normal areas.
Ensure there is a mechanism to take good ideas forward.
Involve the final users of the technology from the start.
Keep the core team small.

The pilot installation was implemented and there was considerable enthusiasm for the technology within the business and it was expected to be used on a much wider basis.

How do we select technology for business benefit?

Process perspective

Managers are commonly faced with difficult decisions about where to invest scarce resources. In the long-term it is critical to select the best technological option, as mistakes can be very costly by the time products and services reach the marketplace. Technology investments must lead to increased future revenues and profits which can be re-invested in the technology base for long-term success.

The key issues in resource prioritization are

What are the decision criteria?
How is a visible and repeatable decision-making process achieved?
What are the strategic implications?
How do we compare with our competitors?

Selection of technology is a decision-making process (see Figure 5.5). It requires an understanding of the technology requirements of the organization, product, service or project, together with the characteristics of identified candidate technologies and any constraints that may affect the selection process. Technology selection involves developing and evaluating

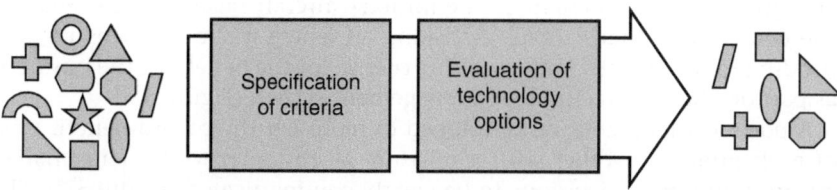

Figure 5.5 Technology selection process

alternative solutions, choosing the best alternative, and considering the most significant implementation factors.

Technology selection decisions can be categorized into two types:

Proactive selection decisions, as a response to future needs:

> Technology forecasting (future investment in technology for the next generation of products).
> Technology portfolio analysis (current / future balance of technology for the near future).

Reactive decision-making, as a response to specific current needs:

> Current investments (project selection).
> Urgent problem-solving (trouble-shooting).

Tackling these issues in a practical way can be difficult. Companies can often identify a set of feasible options but find it difficult to prioritize objectively, especially if there are important qualitative factors and scarce resources.

Case study

This case shows how an aerospace company prioritizes its research and development programs, as part of an overall technology management system.

R&D within the military aircraft sector of industry is extremely diverse, ranging from short-term demands to satisfy new operational requirements to very long-term ones to meet future defense needs. In the current environment, the company (like many other firms) cannot resource all the R&D that the business demands. It therefore looks towards innovative ways of acquiring the technology that it needs through a mixture of contracts, collaborations, and partnerships with both industry and academia, in addition to its own internal R&D programs.

The company identified a need to develop an optimum process for the relative valuation of R&D, to enable selection and prioritization of programs and give maximum benefit to the military aircraft business. This enabled the company to make robust decisions on where it should focus its own funding for R&D, both long- and short-term, for the benefit of the business, as part of a broader technology management system (Figure 5.6).

A portfolio approach was developed to represent the cost-to-benefit ratio of each project, together with a measure of customer focus. This enables resource allocation decisions to be clearly communicated (Figure 5.7). The approach has been successfully applied for several years, leading to an internal business improvement award for the approach.

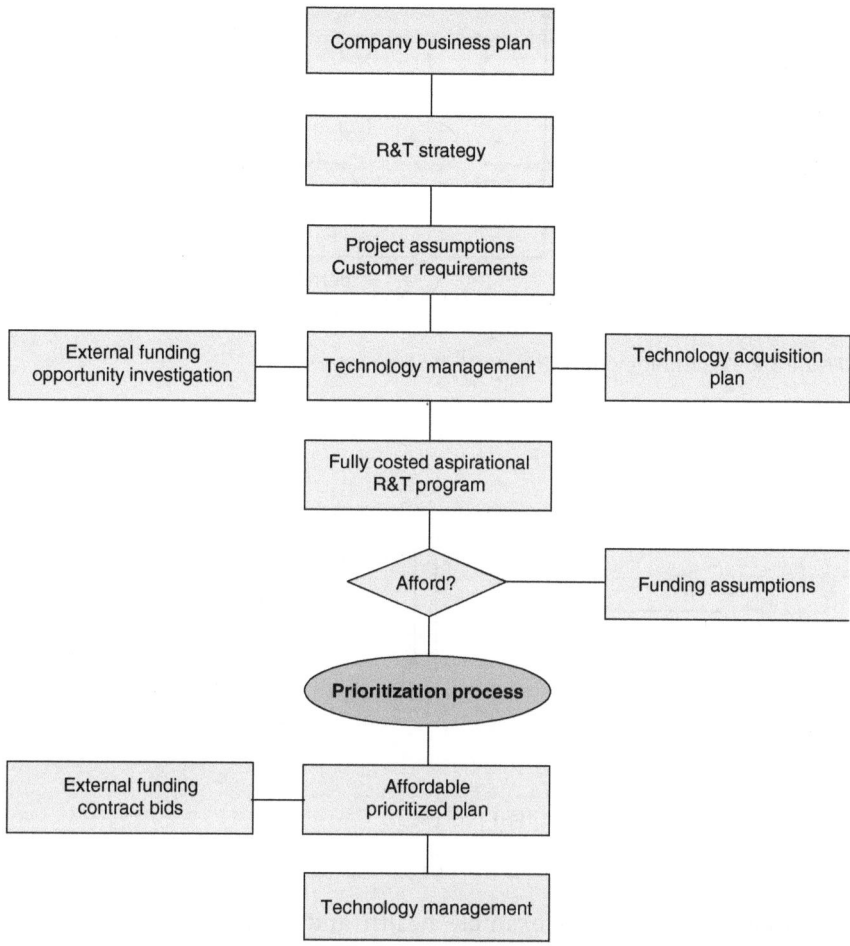

Figure 5.6 Technology management process framework, highlighting selection process

How should we acquire new technology?
Process perspective

Organizations need to update and restock their technology base, which can be depleted by obsolescence and diffusion of technology (Figure 5.8). Specific business reasons for acquiring new technology include:

Customers in a changing market demand new features in products or services.

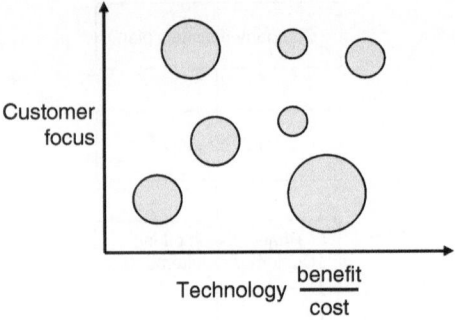

Figure 5.7 Technology selection portfolio framework

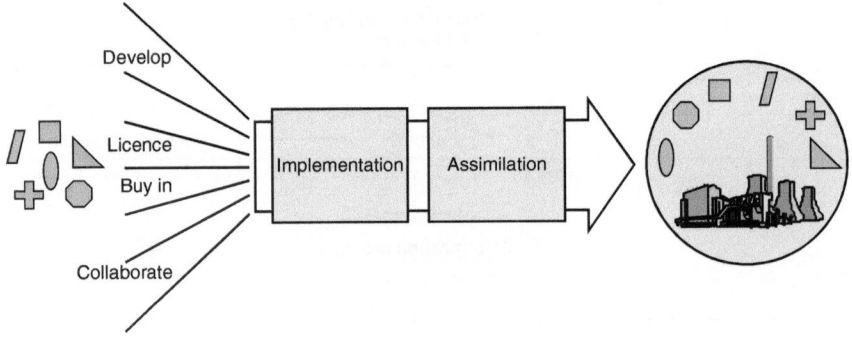

Figure 5.8 Technology acquisition process

External constraints (for example health and safety and environmental legislation) may require the introduction of new products or services.

Increased competition may demand improvements in technological performance.

Improved quality requirements may lead to the necessity for upgrading manufacturing and testing equipment.

Pressures to reduce costs may require more efficient production processes.

Various routes are available for acquiring new technology, including external purchase or transfer of technology (such as company acquisition, machine purchase or licensing in technology), collaborative development (for example, joint ventures, subcontracting development projects, or supporting supplier technology development), and internal technology transfer or R&D.

Technology acquisition can be seen in terms of a general process:

The choice of route for acquiring the technology should be reviewed and assessed.

Implementation of the chosen route should be managed so that the technology is brought into the organization to meet required time, budget, and performance level requirements.

Assimilation of the technology should be achieved to ensure that it becomes a fully accepted and functioning part of the technology base of the company.

Case study

This case summarizes how an industrial inkjet printing company acquired a new technology to develop a new range of products.

Satisfying customer requirements for improved products is a key driver for the acquisition of new technologies in manufacturing companies. The case study company has been at the forefront of inkjet printing technology for marking and coding systems for many years. The requirement from customers for cleaner, more reliable coding technologies encouraged the company to investigate potential alternatives to serve existing markets and to open up new market possibilities.

A systematic review of coding technologies was undertaken, leading to the identification of lasers as a cleaner alternative. Lasers can mark many materials directly, such as plastics or glass, where surface discoloration acts as a mark. To survey the laser marketplace, guidance was sought from a technical consultancy.

The company had a clear strategy for acquiring laser technology. As the laser would be the main differentiating element in a product coding system, the company required complete control over the design and manufacture of the laser. Lasers were too expensive and not sufficiently developed for this application to buy in ready-made. A program of R&D to produce a low-cost, reliable laser for use in a coding system was therefore needed, unless a suitable development partner could be found.

A company in the US was identified which had laser design and manufacturing skills and which had developed a unique, fast, and robust marking product. This company had good laser technology, technologists, and facilities, but had suffered from poor marketing and was not profitable. Such a company would provide the case study company with the technical laser capability it needed to integrate into its next generation marking system. A successful purchase of the company assets was made, ensuring that key technical specialists were retained.

In addition to this, a key customer had identified a small laser marking company in the UK that had developed a high-resolution laser marker. This company was looking for a bigger company to work with in developing its product. The US laser acquisition was made at the same time as developing an exclusive licensing partnership with this UK company. The high-resolution marker was complementary to the US product but could use the same laser technology.

There was initial concern that laser products would compete with existing inkjet technology and replace this element of the business. However, it has been realized that there was a new market waiting in anticipation for the new product and that, far from competing, both technologies have complemented each other.

How can we protect our technology base?

Process perspective

A key part of technology management is the maintenance of the technology base. In addition to ensuring that technological resources are renewed, it is important to minimize unplanned, transfer of technological assets out of the organization. Protection of technology involves more than just patents and intellectual property rights – it involves people and the knowledge and skills they control, together with other issues such as site and computer security.

Key questions include

Do you consider the protectability of the new technologies you are taking on? Are these issues included in the criteria for the selection of technology?

Do you actively manage your technology base so that you are aware of obsolescence or the need for renewal?

Do you monitor and manage the technological expertise of staff and have appropriate reward systems in place to minimize the risks associated with staff turnover?

Technology protection can involve keeping ahead of competitors by identifying and appropriately securing your technology assets (defensive strategy), or by keeping competitors behind by neutralizing the effects of their defenses (proactive strategy).

Protection of technology should be considered systematically, in terms of an ongoing process (Figure 5.9). The three main stages are

Assessment of protection need, including a review of existing and new technology assets in terms of their value to the company (now and in the future).

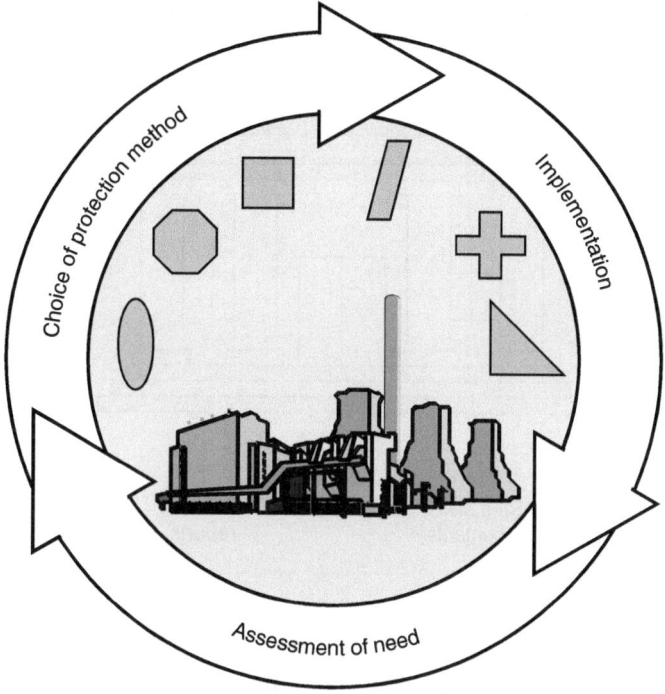

Figure 5.9 Technology protection process

Choice of protection routes, based on their suitability for the technology
   and company.
Implementation and enforcement of the protection method.

Case study

This case study illustrates how technological knowledge is managed in a
large gas supply company.

Organizations such as this case study company are recognizing that a significant part of their value lies in the knowledge that the company and its
employees possess, rather than just in its physical assets. This is particularly
true regarding knowledge of technology. The effective management of that
knowledge can lead to an enhancement in the performance of a company.
However, this can only be achieved by a change in culture and working
methods, whereby the creation and sharing of knowledge is both encouraged and rewarded.

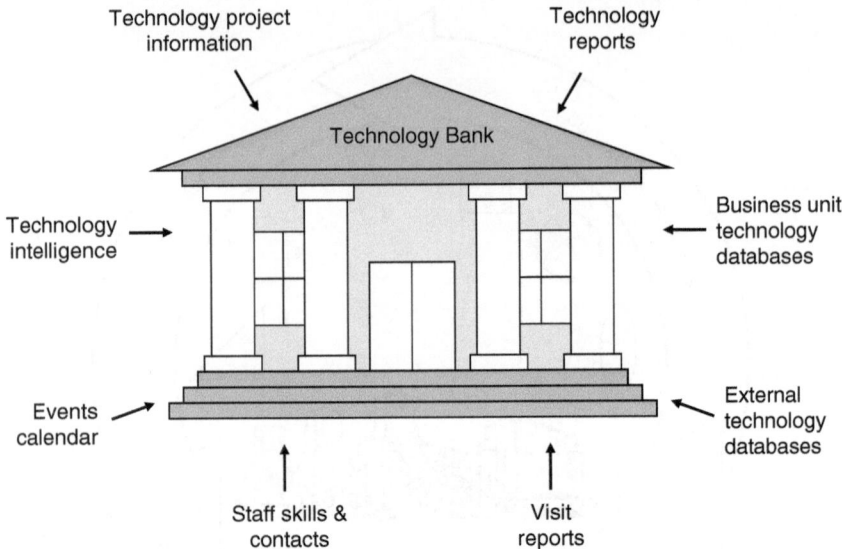

Figure 5.10   Technology Bank

The mission of the Knowledge Management Technologies team within the company is to contribute to and to facilitate the optimization of the use of their world-leading knowledge base in all aspects of gas technology. The approach has been to work with the company businesses to define where knowledge sharing, particularly in the field of technology, can enhance their performance, and then to design and deliver information systems which can realize that potential enhancement.

The development of a Technology Bank (Figure 5.10) is an example of such an activity. The Technology Bank is maintained by Program Managers within the company's Technology group. Current and past technology knowledge is captured in databases within the bank. This then provides the company with the ability to share this knowledge worldwide between virtual teams.

Highly advanced tools have also been developed to search for information simultaneously across all of these and many other databases. Information in the databases can be made visible to everybody within the company or only to restricted groups of people, depending on its level of confidentiality. It is planned to extend this capability to the sharing of selected information with partners outside the company.

This project moved from a development into a pilot phase. The pilot involved users from around the world and will also involve changing the way of working within the Technology group, so that as project information is created it is automatically captured within the databases. The project is just one example of the company's move towards becoming a knowledge-based company, enabling all the knowledge to be accessed quickly and easily from anywhere in the world.

Managerial implications

- There are five generic technology management processes which provide a framework for understanding how technology management activities relate to other business processes and to each other (Figure 5.11):
  - Identification of technologies that are (or may become) of importance to the business.
  - Selection of technologies that should be supported by the organization.
  - Acquisition and assimilation of selected technologies.
  - Exploitation of technologies to generate profit, or other benefits.
  - Protection of knowledge and expertise embedded in products and services.
- Effective technology management requires an integrated approach:
  - Technology should be considered in the early stages of strategy formulation and the links with other activities should be clearly understood (that is marketing and other commercial functions, operations, human resources, and finance).
  - Mechanisms should be in place to ensure that technology strategies are effectively implemented at the operational level.
  - Technology management processes are often embedded in other business processes. For instance, new technology is often acquired during development projects. It is important to be aware of technology

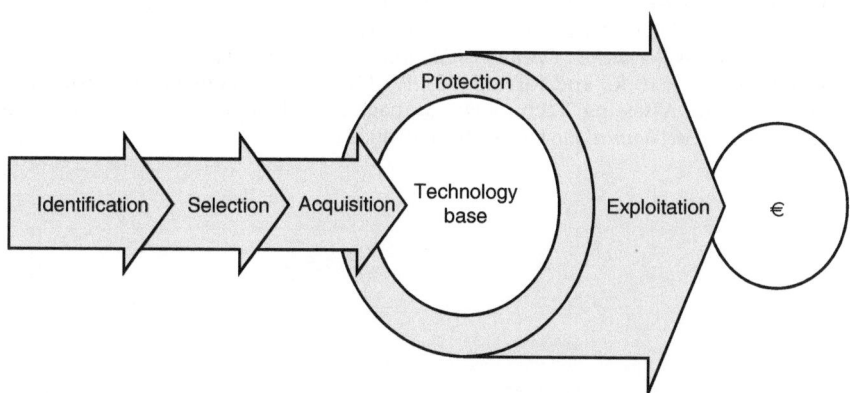

Figure 5.11 Technology management process framework

management considerations that continue beyond the completion of the project.

- The interdependence of technology management processes should be understood, for instance, issues of technology protection are an important consideration during technology identification and selection processes.
- Technology management is a very broad subject, and there are many particular areas of interest that require specific attention. For instance, issues of innovation, knowledge management, competence, and performance measurement are large subjects in themselves. The five-process technology management model provides a framework for understanding how these issues are linked.

References

Cetindamar, D., Phaal, R., and Probert, D. (2010). *Technology Management – Activities and Tools*. Basingstoke: Palgrave Macmillan.

Farrukh, C.J.P., Phaal, R., and Probert, D.R. (2000). *Technology Management Assessment Procedure – a Guide for Supporting Technology Management in Business*. London: Institution of Electrical Engineers.

Farrukh, C., Fraser, P., Hadjidakis, D. Phaal, R., Probert, D., and Tainsh, D. (2004). "Developing an Integrated Technology Management Process." *Research Technology Management* 47(4): 39–46.

Farrukh, C.J.P., Phaal, R., Probert, D.R. Gregory, M., and Wright, J. (2000). "Developing a Process for the Relative Valuation of R&D Programmes." *R&D Management* 30(1): 43–53.

Gregory, M.J. (1995). "Technology Management – a Process Approach." *Proceedings of the Institution of Mechanical Engineers* 209 (Part B): 347–56.

Phaal, R., Farrukh, C.J.P., and Probert, D.R. (2004). "A Framework for Supporting the Management of Technological Knowledge." *International Journal of Technology Management* 27(1): 1–15.

Phaal, R. and Farrukh, C.J.P. (2001). "Technology Management Process Assessment: a Case Study." *International Journal of Operations and Production Management (Special Issue on Process Research in Operations Management)* 21 (8): 1116–32.

Probert, D.R., Phaal, R., and Farrukh, C.J.P. (2000). "Development of a Structured Approach to Assessing Technology Management Practice." *Proceedings of the Institution of Mechanical Engineers* 214 (Part B): 313–21.

6

Successful New Product Development by Optimizing Development Process Effectiveness in Highly Regulated Sectors: The Case of the Spanish Medical Devices Sector

Annemien J.J. Pullen, Carmen Cabello-Medina,
Petra C. de Weerd-Nederhof, Klaasjan Visscher, and Aard J. Groen

Introduction

Innovation is a key driver of sustainable competitive advantage and one of the key challenges for small- and medium-sized companies (SMEs) (O'Regan et al., 2006). Therefore, SMEs need to remain active in new product development (NPD). It is difficult for SMEs[1] in regulated sectors to development new products, because heavy regulatory involvement imposes a number of difficulties on the NPD process. Products have to meet these strict regulations in terms of quality, safety, functionality, and manufacturability, which makes it difficult for SMEs to differentiate in terms of the effectiveness of the product concepts. However, there are big differences in the NPD performance of SMEs. Then, the questions are (1) how do SMEs in regulated sectors distinguish themselves in terms of innovation performance? And, (2) how can SMEs in regulated sectors be successful in NPD?

During this research, the Spanish medical devices sector is used as an illustration of a highly regulated sector. The medical devices development process is characterized by a heavy regulatory involvement (Shaw, 1998). Companies in the medical devices sector are experiencing a need to develop new products more rapidly to satisfy expanding and changing customer requirements in light of new technologies and intensifying global competition (Millson and Wilemon, 2000). The ability of organizations in the medical devices sector to develop and commercialize new products fast is a major competitive advantage (Atun et al., 2002), as speed is an important driver for NPD performance (Lynn et al., 1999; Calantone and Di Benedetto, 2002; Takayama et al., 2002; Langerak and Hultink, 2005).

It is important to realize that in highly regulated sectors, such as the medical development sector, the product concept effectiveness of all acting companies almost per definition will be high, and variance in this performance measure will be low. This is so because all (new) product (concepts) have to comply with the same strict regulations. In these types of sectors, and especially for the SMEs in it, the effectiveness of the NPD process effectiveness stands a much better chance to make a difference. The development process effectiveness represent a measurement of the current NPD performance beyond the requirements imposed by regulations of the sector. This means that it is to be expected that the SMEs we looked at in the Spanish medical devices sector would try to achieve competitive advantage in terms of speed, productivity, and flexibility of their product development process, rather than in terms of manufacturability, functionality, and cost of the product concept, which would be comparable for all players in the field.

According to DeWeerd-Nederhof et al. (2008) both the current and future NPD performance are heavily influenced by the way the NPD function is organized (that is the NPD configuration). The organization of the NPD function consists of the strategy, structure, climate, and process of the NPD function (DeWeerd-Nederhof et al., 2007, 2008). Building on this, and in light of the peculiarities faced by SMEs in highly regulated sectors, we set out to search for a shared pattern in the organization of the NPD function of Spanish SMEs in the medical devices sector, which can be related to high NPD process effectiveness, and ultimately to outperforming competitors.

Thus, our main research goals are first to explore differences in product concept effectiveness and development process effectiveness among SMEs in the Spanish medical devices sector, to see whether or not the current NPD performance would indeed be mainly influenced by the development process effectiveness; and second, to explore whether a shared pattern in the organization of the NPD function can be recognized to affect current NPD performance positively.

In the next section we first provide the theoretical framework on both the current NPD performance, and the variables that are included in the organizational configuration of the NPD function (NPD strategy, structure, climate, and process (DeWeerd-Nederhof et al., 2007). Next we provide the research design and methodology. We then present the research results based on a structured survey among 11 SMEs in the Spanish medical devices sector. The results are further illustrated by two real-life case descriptions. In the discussion and conclusion results are further elaborated and managerial implications are explicitly addressed.

Theoretical framework

NPD performance

The NPD performance consists of the product concept effectiveness on the one hand, and the development process effectiveness on the other hand. The product concept effectiveness is used to define how well a new product concept fits with internal and external characteristics of the company. Whereas the development process effectiveness concept is used to define how effective the development process is executed (Brown and Eisenhardt, 1995). Figure 6.1 shows a schematic overview of the different constructs that together build NPD performance.

The NPD performance is a dynamic concept that has both a short-term and a long-term component. The short-term component is the Operational Effectiveness and refers to the effectiveness of today's work, whereas

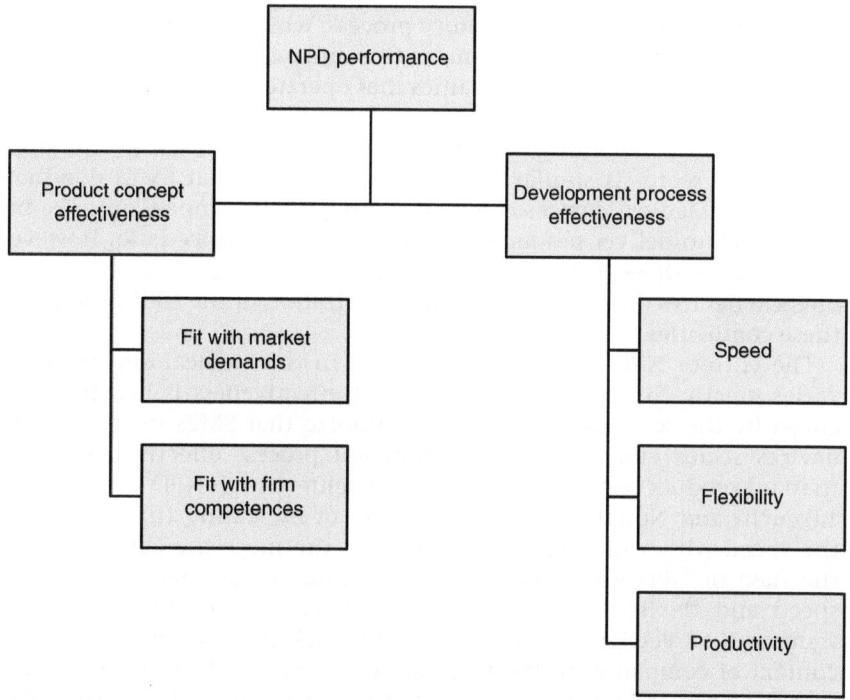

Figure 6.1 Schematic overview of the constructs that together build NPD performance

the long-term component is the Strategic Flexibility which relates to the readiness to adapt to, anticipate or even create future NPD performance requirements (DeWeerd-Nederhof et al., 2008). For this research the focus is on operational effectiveness as the aim is to measure the current NPD performance.

Pettigrew and Whipp (1991) studied organizational change and differences in NPD performance similarly using a content-context-process framework. Content represents the objectives, purpose, and goals of the organization (Pettigrew and Whipp, 1991). Context represents the environment of the company, and process represents the product development process of the organization. The content and context dimensions of Pettigrew and Whipp (1991) can be linked to the product concept effectiveness of Brown and Eisenhardt (1995), whereas the process that Pettigrew and Whipp (1991) describe is similar to the development process effectiveness that Brown and Eisenhardt (1995) describe.

Strict regulations are a unique characteristic of the medical devices sector, and this heavy regulatory involvement characterizes the medical devices development process (Shaw, 1998). The product concept effectiveness is highly tied to this regulatory process, which makes it difficult for companies to differentiate in terms of this dimension. Also, Pettigrew and Whipp (1991) suggest that companies that operate in the same sector (like in the medical devices sector) share environmental characteristics such as regulations, dynamism, and fragmentation of the sector. The medical devices sector is similar to other industries in that SMEs dominate the sector. Medical devices companies often don't compete on price but rather seek to deliver products with a good quality/price-ratio. However, the processes these companies use to achieve their goals and develop new medical devices do differ as does the organization of the NPD function of these companies.

The current NPD performance of SMEs in the medical devices sector varies greatly. Since the product concept effectiveness is heavily influenced by the set regulations, we hypothesize that SMEs in the medical devices sector emphasize on development process effectiveness rather than on product concept effectiveness to achieve high NPD performance. Takeuchi and Nonaka (1986) point this out by stating that the higher the speed with which changes occur and the more the competence in the field of NPD grows, the more firms must focus their processes on speed and flexibility (Takeuchi and Nonaka, 1986). Furthermore the framework of Pettigrew and Whipp (1991) indicates that the content and context of companies in the medical devices sector does not differ, and that they can only distinguish themselves in terms of the process. This supports our previous assumption that companies can distinguish themselves more by focusing on development process effectiveness rather than through product concept effectiveness, and leads to the investigation of our proposition:

Proposition 1: SMEs in the medical devices sector focus on their development process effectiveness rather than on their product concept effectiveness to achieve high NPD performance.

Our study is focussing on the importance of development process effectiveness as part of the current NPD performance. The NPD performance is influenced by the way the NPD function is organized, also called the NPD configuration. Contributing to sustained competitive advantage requires a fit of the NPD configuration with the NPD system and between the NPD system and its context (DeWeerd-Nederhof et al., 2007). The way the NPD function is organized affects both the development process effectiveness, and (to a lesser extent as we proposed) the product concept effectiveness. Differences in development process effectiveness therefore might be explained by the difference in NPD configuration. This leads to the second proposition.

Proposition 2: SMEs in the medical devices sector that achieve high development process effectiveness share a pattern in the organization of their NPD function.

We utilize the concepts of NPD strategy, NPD structure, and NPD climate to further specify the organization of the NPD function (DeWeerd-Nederhof et al., 2007). These concepts are further explained in the following subsections.

NPD strategy

The NPD strategy of a firm can be defined as: "the aggregate pattern of product introductions that emerge from the firm over time" (Firth and Narayanan, 1996). The purpose of the new product strategy is to link the products to the overall objectives of the firm and to assist in the search for new products (Firth and Narayanan, 1996). SMEs with a clear strategy perform better than SMEs that lack a clear strategy (Kargar and Parnell, 1996; O'Regan et al., 2006). Clark and Wheelwright (1993) identify three orientations of the strategy; the technology strategy, the product strategy, and the market strategy. The technology strategy refers to the acquiring, developing, and applying of technology for competitive advantage. The product strategy should contain a clear plan for the development of future products. Finally the market strategy should focus on the question what the target customers will be (Clark and Wheelwright, 1993).

Gatignon and Xuereb (1997) propose a similar typology of strategic orientation (technology orientation, competitive orientation, and customer orientation), and link this to the demand uncertainty in the market. In the medical devices sector the hospital budgets heavily influence the buying behavior of the customers. This buying behavior is also strongly influenced by informal communication between buyers. This causes demand uncertainty in the medical devices sector (Biemans, 1989). When demand uncertainty is high the strategic orientation should be a customer orientation (Gatignon and Xuereb, 1997). In the field of NPD in Spanish firms, Varela and Benito (2005) find that firms that are market oriented get better NPD results than those that do not use this strategic orientation.

Next to the strategic orientation, the project portfolio is an important part of the NPD strategy (Wheelwright and Clark, 1992). Wheelwright and Clark (1992) view NPD strategy as the project portfolio of an organization. It must be clear which type projects are present in the organization. Wheelwright and Clark (1992) distinguish between incremental projects (derivative projects), radical projects (breakthrough projects), and platform projects (between incremental and radical projects) (Wheelwright and Clark, 1992; Gatignon et al., 2002). Incremental innovation projects range from cost-reduced versions of existing products to add-ons or enhancements for and existing production process (Wheelwright and Clark, 1992). Radical innovation projects involve significant changes to existing products and processes. It involves the development or application of significant new technologies or products to markets that are either nonexistent or require dramatic behavior changes to existing markets (Wheelwright and Clark, 1992; Feller et al., 2006).

NPD climate

The second aspect of the organization of the NPD function is the NPD climate. The climate is regarded as a conglomerate of attitudes, feelings, and behaviors which characterizes life in the organization, and exists independently of the perceptions and understandings of the members of the organization (Ekvall, 1996). In order to operationalize climate we use the ten climate dimensions of Ekvall (1996) that stimulate the NPD performance. Cabra (1996) found problems with the challenge dimension by conducting factor analysis with North American samples. Later Isaksen and Lauer (2002) found that the dynamism dimension was not discriminating. In this research we use the dimensions proposed by Ekvall (1996), excluding the dynamism dimension. In this research, a climate that stimulates innovation (innovative climate) is a climate with high levels of "challenge, freedom, idea support, trust, playfulness, debates, risk taking, and idea time" and a low level of conflicts.

NPD structure

The third concept of the organization of the NPD function is the structure of the NPD function. This structure refers broadly to the structure of project teams and the way the people in the NPD function are organized. This work is based on efforts of (Clark and Wheelwright, 1992) who showed that effective product and process development requires teams that integrate people with multiple specialized capabilities. These teams are also referred to as cross-functional product development teams. Cross-functional development teams have become increasingly important due to complexities in the pace, diffusion, and the use of multiple technologies to solve customer problems (Walsh and Linton, 2001) as well as burgeoning global competition (McDonough III, 2000). This is also in line with the research of Sosa et al. (2004) who state that complex product development requires structuring the organization into groups of cross-functional design teams to design systems and components, and with the research of Cooper et al. (2004) who have identified the presence of cross-functional teams as a common fact in organizations they rated as best performers.

Clark and Wheelwright (1992) have characterized a number of structures for project teams. It depends on the environment, organization size, and innovation type which project structure is best suitable (Clark and Wheelwright, 1992). They distinguish between the functional, lightweight, heavyweight and autonomous team structure. The team structure that is used by the company needs to fit in the context. For new product development in the medical devices sector it is very important that all functional areas are involved in the development of a new product, because of the rapid changes in technology and competition. However it should be prevented that a project team gets carried away by its own ideas and fails to meet regulations, or that senior management looses the control over the team (which is likely to occur in the autonomous team structure). Therefore we expect that the heavyweight team structure is likely to be the most successful in the context of the medical devices sector. Also the success factors for cross-functional teams (McDonough III, 2000) can be found most clearly in the characteristics of the heavyweight team structure.

Another aspect of the NPD structure is the formalization of the development process (Griffin and Page, 1993). Formalization refers to the degree in which the process is subject to rules, procedures, and structures previously specified (Johne, 1984). Walsh and Dewar (1987) link the degree of formalization with the organizational life cycle. They state that the more mature the organization, the more formalized the processes are (Walsh and Dewar, 1987). For NPD, it is stated that companies with a formal development process are more successful in the commercialization of new products (Booz and Hamilton, 1982).

We investigate both propositions based on the above literature. The next section describes the methodology we follow to: (1) investigate if SMEs in the Spanish medical devices sector should focus on development process effectiveness to achieve high innovation performance, (2) explore if there is a pattern in the organization of the NPD function that these companies share, and (3) what this organization of the NPD function looks like.

Methodology

We utilize a case based method as described by Yin (2003) and Eisenhardt (1989). We leveraged the international Patterns in NPD project. This project is aimed at developing knowledge in the NPD area, by describing, exploring, and analyzing the organization of the innovation journey. We focus on the population of Spanish SMEs in the medical devices sector.

Sampling process

Consistent with the case study method, we gathered data of a full population in one specific sector, to reduce extraneous variation (Eisenhardt, 1989). Data was gathered in the Spanish medical devices and disposables sector. The medical devices sector is the focus of this research because (1) differences in innovation performance of the companies depend (due to strict regulations) on management issues, and not on environmental or product concept issues, and (2) innovative capability is in this sector of vital importance (Atun et al. 2002). Data gathering took place in the spring of 2006.

Leveraging the Data Universal Numbering System (DUNS) database we used the Spanish SIC codes (CNAE) 33100 and 33200 to identify a number of organizations. A total of 109 companies were selected. These companies were first screened on origin and number of employees. The companies with CNAE 33200 were also screened on the fact whether they were active in the medical devices sector or not. Companies with other origins than Spanish, organizations with a total number of employees of five or less, and organizations (with CNAE 33200) not active in the medical devices sector were deleted from the list. Fifty-seven companies remained and were contacted to find out whether they had an NPD function. From this 35 companies remained, of which 31 companies were interested in participating in the study.

Data description

To the NPD managers of the 31 companies that were interested in participating, a questionnaire about the organization and performance of the NPD function was sent. This questionnaire was developed as part of the international "Patterns in NPD project." We ended up with 12 completed questionnaires from companies in the Spanish medical devices sector, which resulted in a response rate of 34,29 percent.

One of the cases was deleted from the sample, as the number of Full-time Equivalents (FTE) of the particular company was 650 FTE whereas the focus of this research is on small- and medium-sized companies (FTE $\leq$ 250). This resulted in a dataset of N=11 companies, with which the propositions were further explored. Table 6.1 gives general information about the companies in the dataset.

Table 6.1 General information of the companies in the dataset

Case#	FTE	Products	Profit	Sales	Profit/FTE	Sales/FTE
1	12	Interventional cardiology products	€ 113.004,31	€ 4.000.000,00	€ 9.417,03	€ 333.333,33
2	120	Products based in three main lines: Infusion, Respiratory and Bandages	€ 1.064.975,42	€ 10.000.000,00	€ 8.874,80	€ 83.333,33
3	35	Solariums, professional equipment of aesthetic and aesthetic medicine	€ 124.263,70	€ 7.000.000,00	€ 3.550,39	€ 200.000,00
4	80	Four product groups: measuring, quality and metering, industrial electric protection, and power factor protection	€ 5.334.493,30	€ 80.000.000,00	€ 66.681,17	€ 1.000.000,00
5	36	Prostheses and implants	€ 260.478,55	€ 5.000.000,00	€ 7.235,52	€ 138.888,89
6	32	Female protection slips, female hygienic bandages, and children's diapers	€ 44.271,19	€ 3.200.000,00	€ 1.383,47	€ 100.000,00
7	12	Orthopedic elastic products	€ 64.910,78	€ 4.000.000,00	€ 5.409,23	€ 333.333,33
8	167	Medical disposables for neurosurgery and endosurgery	€ 54.000.000,00	€ 54.000.000,00	€ –	€ 323.353,29
9	80	Wide range of single use and reusable lab ware for chemical, clinical, research, and industrial testing laboratories, swabs for sample collection and transport of microbiological material, sampling containers, blood collection tubes, test tubes	€ 72.800,98	€ 18.000.000,00	€ 910,01	€ 225.000,00
10	49	Dental equipment and optical units	€ 683.275,34	€ 9.600.000,00	€ 13.944,39	€ 195.918,37
11	60	Laboratory equipment	€ 765.986,92	€ 18.000.000,00	€ 12.766,45	€ 300.000,00

Measurements

NPD performance is a dynamic concept. It is divided in the current NPD performance (operational effectiveness) which refers to the effectiveness of today's work, and the future NPD performance (strategic flexibility) which relates to the readiness to adapt to, anticipate or even create future requirements (see also Figure 6.1) (Brown and Eisenhardt, 1995; DeWeerd-Nederhof et al., 2005). This research focuses on the current NPD performance, which consists of the development process effectiveness, and the product concept effectiveness. Table 6.2 shows the constructs and items that together form the product concept effectiveness and the development process effectiveness. Table 6.2 also shows the reliability of the constructs and the literature that was used to build the constructs. All items are measurement on a 7-point Likert scale, ranging from "1 = Not at all achieved" to "7 = Very well achieved."

Current NPD performance is measured by using all the scales of the product concept effectiveness and development process effectiveness. Product concept effectiveness is measured as the average score of the constructs "fit with market demands" and "fit with firm competences." Development process effectiveness is measured as the average of the constructs "speed," "flexibility," and "productivity."

We use the development process effectiveness to determine whether a company is high or low performing. If the development process effectiveness of the company is higher or equals the average development process effectiveness of the dataset (which is 4,5), then the company is "high performing." Else the company is "low performing." Table 6.3 shows the scores on product concept effectiveness, and development process effectiveness of the companies in the dataset. Table 6.3 also shows whether the companies are high or low performing based on the above described method.

The NPD climate was measured by asking the respondents to indicate on a 7-point Likert scale to what extent employees have the freedom to define their own work and to what extent there is time for people to develop unplanned new ideas (Pullen et al., 2009). This measurement of NPD climate is based on work by Isaksen and Lauer (2002), and Ekvall (1996), who developed nine items to measure activities related to the climate of the respondents' NPD function. A climate that stimulates innovation is a climate with high levels of "challenge, freedom, idea support, trust, playfulness, debates, risk taking, and idea time" (Ekvall, 1996).

To measure the variable NPD structure, the team structure types of Clark and Wheelwright (1992) were used. In the survey, respondents were asked to indicate whether they use a functional, lightweight, heavyweight or autonomous team structure (Pullen et al., 2009).

Table 6.2 Overview and reliability statistics of the performance scale

| | Current NPD performance | | | | |
| | Product concept effectiveness (pce) | | Development process effectiveness (NPDpe) | | |
Construct	Fit with market demands	Fit with firm competences	Speed	Flexibility	Productivity
N of items	6	6	6	6	6
Measurement scale	7-point Likert scale	7-point Likert scale	7-point Likert scale	7-point Likert scale	7-point Likert scale
Cronbach's alpha	$\alpha = 0{,}788$	$\alpha = 0{,}747$	$\alpha = 0{,}893$	$\alpha = 0{,}645$	$\alpha = 0{,}778$
Based on	Customer satisfaction, timeliness, product price, quality (Chiesa et al., 1996) Sales and profit impact (Bretani and Kleinschmidt, 2004)	R&D/ Manufacturing integration (Swink 1999; Yam et al., 2004) R&D/Marketing integration (Leenders and Wierenga, 2002)	Speed relative to schedule (Kessler and Bierly, 2002) Development time (DT), concept to customer time (CTC), total time (TT) (Griffin, 1997) Speed and commitment of the NPD decision-making process, (Griffin and Page, 1993)	Average time and cost of redesign, enhancement (Chiesa et al., 1996; Thomke, 1997) The ability to change specs late (Thomke, 1997)	The possibility for lower development budget (Iansiti, 1993) Cost relative to budget, competitors (Kessler and Bierly, 2002) Engineering hours, cost of materials, cost of tooling (Clark and Wheelwright, 1993)

Table 6.3   Performance scores of the companies in the dataset

Case #	Product concept effectiveness	Development process effectiveness	High/ Low performing
1	5,9	2,5	Low
2	4,3	4,0	Low
3	4,0	4,1	Low
4	5,3	4,7	High
5	4,3	3,3	Low
6	4,8	4,5	High
7	5,3	5,3	High
8	6,4	5,1	High
9	5,7	4,9	High
10	6,1	5,9	High
11	4,8	4,7	High
Average	5,2	4,5	

The level of formalization and presence of cross-functional teams was measured by presenting multiple descriptions of development processes of a business unit. Based on descriptions of the NPD system by Griffin and Page (1993), the respondents were asked to indicate which development process most closely describes the development process that is used in their business unit (Pullen et al., 2009).

The strategic orientation was measured with a seven-point Likert scale ranging from "1 = strongly disagree" to "7 = strongly agree." Respondents were asked to indicate the level of agreement with statements considering the technology strategy, product strategy, and market strategy (Clark and Wheelwright, 1993).

To measure a company's NPD portfolio the respondent was asked to indicate the percentage radical, incremental and next generation projects in the portfolio (Wheelwright and Clark, 1992). The percentages had to sum up to 100 percent.

Data analysis techniques

For analysis of the data we first rely on a theoretical proposition (Yin, 1994). We are interested in (1) the variance of both the product concept effectiveness and the development process effectiveness, and (2) the organization of the NPD function that the companies in our dataset possibly share. The variances are calculated in the statistics program SPSS. We conducted a structured survey in 11 SMEs in the medical devices sector. In line with the methodological suggestions of Eisenhardt (1989) we made case summaries

Table 6.4   Single respondent bias results

Test statistics casestudy 1[b]

	NPDmanager_1 Minisurvey_1
Z	–,178a
Asymp. Sig. (2-tailed)	,859

a. Based on positive ranks.
b. Wilcoxon signed ranks test

Test statistics casestudy 2[b]

	NPDmanager_2 Minisurvey_2
Z	–1,244a
Asymp. Sig. (2-tailed)	,214

a. Based on positive ranks.
b. Wilcoxon signed ranks test

and analyzed each case individually. In addition to the structured survey, we conducted 2 in-depth case studies: one in the highest performing company, and one in the one but lowest performing company of our dataset. These case studies (1) show if there is single respondent bias or not (see next paragraph), and (2) give background information and enlighten the results we found with the structured survey.

Single respondent bias

One of the problems of response in survey research is single respondent bias. We compensated this by controlling for single respondent bias. From our dataset of 11 companies we selected two companies for case studies on the climate variable. The companies were selected on their scores on development process effectiveness and current NPD performance (highest scoring company and lowest scoring company). Besides the full questionnaire that was filled in by the NPD manager, at least five employees in both companies filled in a mini survey that was solely focused on the NPD climate. In this way we could compare the filled in answers of the NPD manager to those of different employees in the company. For both cases we found no significant difference (Sign. p> 0.00 for both cases) between the answers of the NPD manager who filled in the full questionnaire and the answers that were given by the employees in the mini surveys (see Table 6.4). This excludes single respondent bias.

Results

We have presented two propositions which we tested. Our first proposition was that SMEs in the medical devices sector focus on their development process effectiveness rather than on their product concept effectiveness to achieve high NPD performance. Table 6.5 shows the results of the variance

Table 6.5   Variances in product concept effectiveness (PCE) and development process effectiveness (NPDpe)

	N	Mean	Std. deviation	Variance
PCE	11	5,1523	,78841	,622
NPDpe_real	11	4,4693	,95286	,908
Valid N (listwise)	11			

in both the product concept effectiveness and the development process effectiveness.

We calculated the variances in both concepts to see whether the scores of the product concept effectiveness indeed vary less, or are more stable, than the scores of the development process effectiveness. When the variance of the product concept effectiveness between the companies is low, the product concept effectiveness is not the construct that makes it possible for companies to distinguish themselves in terms of current NPD performance. Instead, the development process effectiveness is what distinguished companies in terms of current NPD performance. It becomes apparent from Table 6.5 that (1) the variance of the product concept effectiveness is low and (2) the variance of the development process effectiveness is high (see Table 6.5). This indicates that the development process effectiveness is indeed the variable that distinguishes between the NPD performances of SMEs.

Our second proposition was that SMEs in the medical devices sector that achieve high development process effectiveness share a pattern in the organization of their NPD function. We divided the dataset in high and low performing companies based on the standards described and shown in Table 6.3 in the measurements section. The case summaries in Table 6.6 show the organizational patterns of the NPD functions amongst the high performers and amongst the low performers.

At first glance, the case summaries in Table 6.6 show a lot of variety in the organization of the NPD function. However, when taking a closer look, a number of patterns in the organization of the NPD function become apparent.

NPD strategy

A first pattern can be found in the project portfolio of the companies. The high performing companies focus in general on incremental innovation projects, whereas the low performing companies focus more on radical innovation projects. This might be explained by the highly regulated sectors in which these companies operate. The NPDs must meet fixed standards which leaves little room for radical innovations. It is safer to focus

Table 6.6   Case summaries of the internal organization of the companies in the dataset

Development process effectiveness	Case#	Portfolio	Team_ structure	Formalization	Climate
Low	1	Main focus on radical innovation	Heavyweight team structure	No formalized process	No innovative climate
	5	Main focus on incremental innovation	Heavyweight team structure	No formalized process	No innovative climate
	2	Main focus on radical innovation	Heavyweight team structure	Formalized process	Innovative climate
	3	Main focus on radical innovation	Functional team structure	No formalized process	No innovative climate
High	6	Main focus on radical innovation	Heavyweight team structure	No formalized process	Innovative climate
	11	Main focus on incremental innovation	Functional team structure	No formalized process	Innovative climate
	4	Main focus on incremental innovation	Functional team structure	No formalized process	Innovative climate
	9	Main focus on incremental innovation	Functional team structure	No formalized process	No innovative climate
	8	Main focus on incremental innovation	Heavyweight team structure	No formalized process	Innovative climate
	7	Main focus on radical innovation	Autonomous team structure	Formalized process	No innovative climate
	10	Main focus on incremental innovation	Functional team structure	No formalized process	No innovative climate
Total N	11				

on incremental innovation projects, since these types of projects can easier meet regulations than radical innovation projects.

NPD structure

The second pattern is found in the link between team structure and portfolio. The high performing companies 4, 9, 10, and 11 combine an incremental project portfolio with a functional team structure. These findings suggest that the combination of an incremental project portfolio with a functional team structure leads to high development process effectiveness. This is also in line with the research of De Visser et al. (2009) who find that "firms that manage to apply a cross-functional integration structure for their radical NPD processes and a functional integration structure for their incremental NPD processes will be the most successful in terms of balancing derivative and breakthrough innovation performance" (De Visser et al., 2009).

Furthermore our findings suggest that the combination of a radical project portfolio with a heavyweight or autonomous team structure (as seen in case companies 6 and 7) can also lead to high development process effectiveness, when combined with an informal NPD process and innovative climate, or with a formal NPD process and climate that is not innovative.

NPD climate and NPD process

From the (low performing) case companies 1, 3, and 5 in our dataset, it seems that lacking both a formalized NPD process and an innovative NPD climate doesn't lead to high development process effectiveness, unless combined with a functional team structure like in the high performing case companies 9 and 10. In these two latter cases, the functional team structure compensates the lack of formalization to some extent. Also having both a formalized NPD process and innovative NPD climate, like in case company 2, doesn't lead to high development process effectiveness. Combining a formalized NPD process with a NPD climate that isn't innovative and vice versa, seems to lead to high development process effectiveness. This can be seen in the high performing case companies 4, 6, 7, 8, and 11 and is the third pattern we find.

The above results show that, companies in the Spanish medical devices sector indeed share a pattern in their NPD function. This supports our second proposition. To summarize, we found a number of patterns in the organization of the NPD function of high vs. low performing companies.

First of all, indeed the companies in the dataset which focused on the effectiveness of their development process, stood out in NPD performance. Further, the higher performing companies did have a number of commonalities in the organization of their NPD function:

1. The majority of the higher performing firms had an NPD strategy characterized by a predominantly incremental project portfolio.

2. a) Successful firms with an incremental project portfolio combined this with a functional team structure.
3. b) Successful firms with a radical project portfolio combined this with a heavyweight or autonomous team structure.
4. A negative reciprocal relationship exists between formalization of the NPD processes and the climate of the NPD function, in that a formalized NPD process and an innovative climate do not seem to reinforce each other. Innovative climate combined with an informal NPD process does however contribute positively to NPD performance. This effect was stronger in combination with a radical project portfolio.

What the above summarized research results mean in everyday business practice is illustrated in the following two cases. Both companies are part of our dataset of Spanish medical devices companies. Company 5 is the last but one lowest performing company, Company 10 is the highest performing company.

Case company 5: a low performer

Company 5 is a low performing company that focuses on the development, production, and commercialization of prostheses and implants. They want to offer a complete range of products to their clients (surgeons) even though a number of these products are not profitable. In addition, time is not regarded the most important. Over the years, the company has focused more and more on R&D, and they also work on their image of an innovative company. The role of senior management in this is to set an example to the employees and improve the work where possible. However, employees are not stimulated nor compensated to come up with new ideas or new developments. When employees come up with new ideas, the management listens to the ideas of the employees and approves or disapproves and gives advice about other possibilities. Most of the time these new ideas are shared only among fellow employees, as employees are not stimulated (nor compensated) to come up with innovative ideas or new developments. Conflicts between R&D and commercial functions arise when a time plan and quality are promised to customers which are not feasible in practice. Risk taking in NPD by the employees and the management is low.

The level of risk taking in company 5 is low and, as described in text box 1, the focus is on incremental new products (in line with pattern 1). The focus on incremental innovation projects is combined with a heavyweight team structure in which project teams are to a large extent autonomous and project team leaders have the authority to decide about the division of the budget and people within the project. This type of team structure is more applicable to radical innovation projects, since these projects need more freedom to think "outside-the-box," without being constrained by everyday company boundaries. In incremental innovation projects this heavyweight team structure is often too heavy in that in incremental innovation projects the project team should remain close to everyday company business, without getting carried away. A functional team structure is in the case of incremental innovation better applicable. However company 5 combines a focus on incremental innovation with a heavyweight team structure (conflicts with pattern 2). From text box 1 it becomes clear, that the climate in company 5 is not innovative, since employees are not stimulated nor compensated to come up with new ideas or new developments. Management decides about new product development projects, which are executed in a development process that isn't formalized. This combination of a process that isn't formalized and a climate that isn't innovative conflicts with pattern 3.

Only pattern 1, a focus on incremental innovation projects, can be found in company 5. Neither pattern 2 (the presence of a functional team structure in combination with an incremental product portfolio), nor pattern 3 (the reciprocal relationship between formalization of the NPD process and the climate of the NPD function) are present in company 5. The fact that the majority of the organizational patterns that were found to positively contribute to NPD performance miss in company 5 might explain its low NPD performance.

Case company 10: a high performer

Our second case company, company 10, is a high performing company that focuses on dental equipment and optical units. They offer solutions to other companies (they work for) and increase patient comfort with their products. They want to concentrate on further exploitation of the markets they currently serve, instead of focussing on radically new products. They want to grow, but also stay a medium-sized company. It should be a controlled increase. Part of the products are developed for other companies and part of the products are developed for the market. Meeting the – tight – time schedules is of highest importance. The senior management coordinates all the work and ideas in a functional team structure. Every three months product meetings are organized which people from every department must attend. In these meetings ideas are shared with the management, and are selected. The selected ideas are tested by the technical department and if the idea fits within the current technologies and products it will be further explored. However, the final decisions are made top-down. Risk taking is only accepted if it is in line with current technologies and products.

Case company 10 clearly focuses on incremental innovation projects (in line with pattern 1). Text box 2 explains that company 10 wants to exploit their current market further and new product development projects should fit with current technologies and products. This focus on incremental product development projects is combined with a functional team structure (in line with pattern 2) in which management coordinates all the work. The climate is more innovative than in case company 5, because employees in company 10 have room to discuss their ideas in organized informal product meetings (see text box 2). However the climate in company 10 is not that innovative since only incrementally new ideas are appreciated and final decisions are all made top-down. The go/ no go decision about the development project is formal. However, the development process itself is not formalized. The combination of a development process that is not formalized with a climate that is not innovative is compensated in company 10 through the functional team structure (in line with pattern 3).

The organizational patterns 1, 2, and 3 that were found to contribute positively to NPD performance are all present in case company 10. The fact that all three patterns are present in company 10, and the fact that the majority of these patterns is missing in company 5 might explain the difference in NPD performance between both companies.

Discussion

Our findings raise some questions about the organization of NPD in highly regulated sectors. We find that companies in the highly regulated medical devices sector should focus on incremental innovation projects for high current NPD performance. Does this mean that these companies have to neglect radical innovation projects? The fact that our research findings state that a majority of incremental projects should be present can be explained by our focus on current NPD performance, which reflects the NPD performance on the short term. To be able to also achieve high future (long-term) NPD performance a company should not only be operational effective, but also strategically flexible (DeWeerd-Nederhof et al., 2008). To achieve high future NPD performance the project portfolio should also contain projects that gain future revenues even though they aren't profitable at first glance. This is often the case with radical innovation projects. We expect that when the focus is on future NPD performance, radical innovation projects should be more dominantly present in the project portfolio. When the focus shifts from current to future NPD performance we expect that the organization of the NPD function shifts from an operational effective organization with a focus on incremental innovation projects, to a strategically flexible organization with a focus on radical innovation projects.

With regard to the formalization of the NPD process and innovativeness of the NPD climate, we found a negative reciprocal relationship, in that a formalized NPD process and an innovative climate do not seem to reinforce each other. Innovative climate combined with an informal NPD process does however contribute positively to NPD performance. These findings conflict with theory. On the one hand, theory stated that a climate that stimulates innovation is a climate with high levels of "challenge, freedom, idea support, trust, playfulness, debates, risk taking, and idea time" (Ekvall, 1996). On the other hand, theory states that, companies with a formal development process are more successful in the commercialization of new products (Booz and Hamilton, 1982). Now, is theory wrong, or not applicable? Theory is not wrong and is also applicable, but the theoretical approach towards these variables should be more subtle. Companies do not consist of only one variable or characteristic, but of a multitude of variables and characteristics that are all interrelated.

Finally, we focused on a highly regulated sector and found that companies in this sector can only compete on development process effectiveness. This is caused by the fact that the product concept effectiveness is to a great extent predetermined by the set regulations. The product concept effectiveness of companies in sectors that are not highly regulated is not predetermined, which means that companies in nonregulated sectors have not only the possibility to compete on development process effectiveness, but also on product concept effectiveness. Then, to what extent do our research findings also apply in nonregulated sectors?

The short-term/ long-term effects of the project portfolio on the NPD performance also apply in nonregulated sectors. Incremental innovation projects lead to higher revenues on the short term, whereas radical innovation projects lead to higher revenues on the long term. The other patterns we found (pattern numbers 2 and 3) are strongly related to the achievement of high development process effectiveness. We expect that these patterns also apply in nonregulated sectors. However only increasing the development process effectiveness in companies in nonregulated sectors has probably less effect on the NPD performance as increasing the development process effectiveness in highly regulated companies. In nonregulated sectors, also the differences in product concept effectiveness are heavily influencing the NPD performance and need to be taken into account.

Conclusions

The contribution of the research outlined above is that it shows SMEs in regulated sectors how competitive advantage in terms of NPD performance could be achieved, namely by optimizing their development process effectiveness and by choosing an appropriate organization of the NPD function. The research explicitly focused on the combination of organizational variables instead of focusing only on one variable, which adds value to other scholarly work on the same topic.

In line with our theoretical proposition, we find that small- and medium-sized companies in the Spanish medical devices sector can indeed improve the performance of their NPD function by focusing on the speed, flexibility, and productivity of their NPD function. Furthermore we find that, companies with high current NPD performance in terms of development process effectiveness have a number of commonalities in the organization of their NPD function. These companies either combine an incremental project portfolio with a functional team structure, or they combine a radical project portfolio with a heavyweight or autonomous team structure. It should be noted that most of the firms with high development process effectiveness employed an NPD strategy focusing on incremental innovation. Further, a reciprocal relationship between formalization of the NPD processes and the climate of the NPD function was found, in that a formalized NPD process and an innovative climate do not seem to reinforce each other. Innovative climate combined with an informal NPD process does however contribute positively to NPD performance, especially for the minority of firms in the set with an NPD strategy focusing more on radical innovation.

It should be noted however, that as was explained in the theoretical framework section, the NPD performance is a dynamic concept that has both a short-term (Operational Effectiveness) and a long-term (Strategic Flexibility) component. For this research the focus is on operational effectiveness as the aim is to measure the current NPD performance. Although the results of our study might lead one to believe that in highly regulated sectors the only way to innovate is in incremental steps, this is somewhat misleading because of the short-term operational effectiveness view employed in the research. For radical innovation to lead to competitive advantage some organizational characteristics also have been found, but the beneficial effect on both development process and product concept effectiveness might be subject to considerable time delay, especially in the medical devices sector.

For further research we strive to conduct longitudinal research in this field. The data of this research was gathered at one point in time, but since NPD is dynamic, longitudinal research might be interesting. Furthermore, it could be worthwhile to test our research findings in other countries and other strictly regulated sectors. We specifically looked at the context of the Spanish medical devices sector, but since the strict regulations for new medical devices are comparable in most countries, our findings might be applicable in other countries. Also, there are a number of other sectors that have similar characteristics in terms of regulations. Although further research is needed, we expect to find a similar pattern in the internal organization of the NPD function of successful companies in other highly regulated sectors for a larger dataset. Suggestions for other sectors are the biotechnology (Senker, 1991) and commercial space sector (Carayannis and Samanta Roy, 2000).

Managerial implications

So, what do the research findings mean in everyday business practice? It's not possible to give a full recipe for successful NPD, but we can demonstrate the value of certain ingredients, and, just as importantly, warn for the excessive use of some other ingredients. There are several myths about the organization of NPD that are among CTOs and managers of NPD. In this research we tackled four of these myths.

- Myth 1: First, focus on the quality, safety, and manufacturability of the product, then take a look at your NPD process.

We have shown that, in a regulated sector, the quality, safety, and manufacturability standards are predetermined through regulations. High quality, safety, and manufacturability of products are a precondition, regardless of the company, and not leading to competitive advantage. As a manager, you should focus on your NPD process. The development speed should be high (don't waste time), the development process should be flexible (be able to change fast if specifications change), and the development process should have high productivity (don't exceed costs nor budgeted hours).

- Myth 2: The more innovative, the better.

Managers are often confronted with the idea that radical innovation is just it. We have shown that taking little steps in the innovativeness of new products is – at least in regulated sectors – more successful. Managers should take a look at the portfolio of different innovation projects in their companies. How is the balance between incremental and radical innovation projects? If the portfolio mainly contains radical innovation projects and lacks incremental innovation projects, they should try to shift this balance by attracting more incremental innovation projects. However, keep in mind that the pursuit of radical innovations should not be fully abandoned, since they are needed for future profits.

- Myth 3: Project teams should be autonomous and not restricted by organizational procedures.

There is not one best way to structure your NPD teams. The best way to organize projects heavily depends on the type of development projects. As a manager you should take a look at your project portfolio and at the team structure you use. In an incremental project portfolio, the projects are not so new and unknown that you need self-steering project teams. Rather, project teams are required that remain close to the company and do not get carried away. For incremental innovation, you should create project teams in which members remain on their current locations, in which different functions coordinate ideas through detailed specifications, in which occasional meetings are organized to discuss issues that cut across groups, and in which the responsibility passes sequentially from one function to the next. The more radically new the project is, the more the final project responsibility shifts towards the project leader and the more responsibilities the project team should get in general.

- Myth 4: The NPD climate should be innovative and the NPD process should be formal.

We have shown that the innovativeness of the climate and the formalization of the NPD process do not reinforce each other. It is either-or, not both. This means that, there are two roads to success: you, as a manager, either work on an innovative climate, or you work on a well formalized NPD process. Considering the NPD climate, questions you need to pose to yourself are: how much time, freedom, support, and trust do employees get to develop new ideas? Are employees challenged? Are employees allowed to take risks? If you answer most of these questions positively, the climate in your NPD function can be considered innovative. If you answer most of these questions negatively, you haven't got an innovative climate. Considering the formalization of the NPD process ask yourself if your organization follows a formally documented NPD process or not. For high current NPD performance either an innovative climate or a formalized process should be present.

Acknowledgement

The research described in this article is part of the European research project "Patterns in New Product Development: consistent NPD configurations for sustained innovation," which was founded at the University of Twente, the Netherlands. "Patterns in NPD" is funded by RADMA and the University of Twente (The Netherlands).

The case studies were conducted thanks to the enthusiastic cooperation of the NPD managers, CEOs, and employees of two Spanish medical devices companies in the Barcelona and Valencia region in Spain.

Note

1 According to European standards, SMEs are defined as companies that have 250 or less full time equivalents, Commission of the European Communities (2003),

"Commission Recommendation of 6 May 2003 Concerning the Definition of Micro, Small and Medium-Sized Enterprises *(notified under document number C (2003) 1422)* 2003/362/EC." *Official Journal of the European Union* 46 (L124):

References

Atun, R., Shah, S., and Bosanquet, N. (2002). "The Medical Devices Sector: Coming Out of the Shadow." *European Business Journal* 14: 63–72.

Biemans, W. G. (1989). *Developing Innovations within Networks: With An Application to the Dutch Medical Equipment Industry.* Doctoral Thesis: University of Eindhoven.

Booz, Allen and Hamilton Inc. (1982). *New Products Management for the 1980s.* New York: Booz, Allen and Hamilton.

Bretani, U. de and Kleinschmidt, E. J. (2004). "Corporate Culture and Commitment: Impact on Performance of International New Product Development Programs." *Journal of Product Innovation Management* 21(5): 309–33.

Brown, S. L. and Eisenhardt, K. M. (1995). "Product Development: Past Research, Present Findings, and Future Directions." *Academy of Management Review* 20(2): 343–78.

Cabra J. F. (1996). "Examining the Reliability and Factor Structure of the Climate for Innovation Questionnaire." Unpublished Master's Thesis, State University College: Buffalo State University.

Calantone, R. J. and Di Benedetto, C. A. (2002). "Performance and Time to Market: Accelerating Cycle Time with Overlapping Stages." *IEEE Transactions on Engineering Management* 47(2): 232–44.

Carayannis, E. G. and Samanta Roy, R. I. (2000). "Davids Vs Goliaths in the Small Satellite Industry: the Role of Technological Innovation Dynamics in Firm Competitiveness." *Technovation* 20: 287–97.

Chiesa, V., Coughlan, P., and Voss, C. A. (1996). "Development of a Technical Innovation Audit." *Journal of Product Innovation Management* 13(2): 105–36.

Clark, K. B. and Wheelwright, S. C. (1992). "Organizing and Leading 'Heavyweight' Development Teams." *California Management Review* (Spring 1992): 9–28.

Clark, K. B. and Wheelwright, S. C. (1993). *Managing New Product and Process Development.* New York: The Free Press.

Commission of the European Communities (2003). "Commission Recommendation of 6 May 2003 Concerning the Definition of Micro, Small and Medium-Sized Enterprises (notified under document number C (2003) 1422) 2003/362/EC." *Official Journal of the European Union* 46 (L124): 36–41.

Cooper, R. G., Edgett, S. J., and Kleinschmidt, E. J. (2004). "Benchmarking Best NPD Practices – I." *Research Technology Management* (January-February 2004): 31–43.

De Visser, M., DeWeerd-Nederhof, P. C., Faems, D., Van Looy, B., and Visscher, K. (2009). "Ambidexterity in NPD: The Impact of Differentiated Integration Structures on Innovation Performance." In Best Paper Proceedings of the Academy of Management Annual Meeting 2009. Chicago, USA.

DeWeerd-Nederhof, P. C., Altena, J., Visscher, K., and Fisscher, O. (2005). "Assessing Operational NPD Effectiveness and Strategic NPD Flexibility for the Design of New Product Development Configurations." In CDRom Proceedings of the Cinet Conference. Sept. 4–6, 2005. Brighton, UK.

DeWeerd-Nederhof, P. C., Bos, G. J., Visscher, K., Gomes, J. F., and Kekale, K. (2007). "Patterns in NPD: Searching for Consistent Configurations. A Study of Dutch, Finnish, and Portuguese Cases." *International Journal of Business Innovation Research* 1(3): 315–36.

DeWeerd-Nederhof, P. C., Visscher, K., Altena, J., and Fisscher, O. A. M. (2008). "Operational Effectiveness and Strategic Flexibility. Scales for performance assessment of New Product Development systems." *International Journal of Technology Management* 44(3/4): 354–72.

Eisenhardt, K. M. (1989). "Building Theory from Case Study Research." *Academy of Management Review* 14(4): 532–50.

Ekvall, G. (1996). "Organizational Climate for Creativity and Innovation." *European Journal of Work and Organizational Psychology* 5(1): 105–23.

Feller, J., Parhankangas, A., and Smeds, R. (2006). "Process Learning in Alliances Developing Radical versus Incremental Innovations: Evidence from the Telecommunications Industry." *Knowledge and Process Management* 13(3): 175–91.

Firth, R. W. and Narayanan, V. K. (1996). "New Product Strategies of Large, Dominant Product Manufacturing Firms: An Exploratory Analysis." *Journal of Product Innovation Management* 31: 334–47.

Gatignon, H., Tushman, M. L., Smith, W., and Anderson, P. (2002). "A Structural Approach to Assessing Innovation: Construct Development of Innovation Locus, Type, and Characteristics." *Management Science* 48(9): 1103–22.

Gatignon, H. and Xuereb, J. M. (1997). "Strategic Orientation of the Firm and New Product Performance." *Journal of Marketing Research* 34: 77–90.

Griffin, A. (1997). "PDMA Research on New Product Development Practices: Updating Trends and Benchmarking Best Practices." *Journal of Product Innovation Management* 14: 429–58.

Griffin, A. and Page, A. L. (1993). "An Interim Report on Measuring Product Development Success and Failure." *Journal of Product Innovation Management* 10(4): 291–308.

Iansiti, M. (1993). "Real-world R&D: Jumping the Product Generation Gap." *Harvard Business Review* 71(3): 138–47.

Isaksen, S. G. and Lauer, K. J. (2002). "The Climate for Creativity and Change in Teams." *Creativity and Innovation Management* 11(1): 74–86.

Johne, F. A. (1984). "How Experienced Product Innovators Organize." *Journal of Product Innovation Management* 1(4): 210–23.

Kargar, J. and Parnell, J. A. (1996). "Strategy Emphasis and Planning Satisfaction in Small Firms: An Empirical Investigation." *Journal of Business Strategies* 13(1): 42–64.

Kessler, E. H. and Bierly, P. E. (2002). "Is Faster Really Better? An Empirical Test of the Implications of Innovation Speed." *IEEE Transactions on Engineering Management* 49 (1): 2–12.

Langerak, F. and Hultink, E. J. (2005). "The Impact of New Product Development. Accelaration Approaches on Speed and Profitability: Lessons for Pioneers and Fast Followers." *IEEE Transactions on Engineering Management* 25(1): 30–42.

Leenders, M. A. and Wierenga, B. (2002). "The Effectiveness of Different Mechanisms for Integrating Marketing and R&D." *Journal of Product Innovation Management* 19(4): 305–17.

Lynn, G. S., Abel, K. D., Valentine, W. S., and Wright, R. C. (1999). "Key Factors in Increasing Speed to Market and Improving New Product Success Rates." *Industrial Marketing Management* 28(4): 319–26.

McDonough III, E. F. (2000). "Investigation of Factors Contributing to the Success of Cross-Functional Teams." *Journal of Product Innovation Management* 17: 221–35.

Millson, M. R. and Wilemon, D. (2000). "Managing Innovation in Medical Devices." *International Journal of Healthcare Technology and Management* 2(5/6): 425–55.

O'Regan, N., Ghobadian, A., and Sims, M. (2006). "Fast Tracking Innovation in Manufacturing SMEs." *Technovation* 26: 251–61.

Pettigrew, A. M. and Whipp, R. (1991). *Managing Change for Competitive Success.* Oxford: Basil Blackwell.

Pullen, A. J. J., DeWeerd-Nederhof, P. C., Groen, A. J., Song, X. M., and Fisscher, O. A. M. (2009). "Successful Patterns of Internal SME Characteristics Leading to High Overall Innovation Performance." *Creativity and Innovation Management* 18(3): 209–23.

Senker, J. (1991). "National Systems of Innovation, Organization, and Learning in Industrial Biotechnology." *Technovation* 16(5): 219–29.

Shaw, B. (1998). "Innovation and New Product Development in the UK Medical Equipment Industry." *International Journal of Technology Management* 15(3–5): 433–45.

Sosa, M. E., Eppinger, S. D., and Rowles, C. M. (2004). "The Misalignment of Product Architecture and Organizational Structure in Complex Product Development." *Management Science* 50(12): 1674–89.

Swink, M. (1999). "Threats to New Product Manufacturability and the Effects of Development Team Integration Processes." *Journal of Operations Management* 17(6): 691–709.

Takayama, M., Watanabe, C., and Griffy-Brown, C. (2002). "Alliance Strategy as a Competitive Strategy for Successively Creative New Product Development: The Proof of the Co-Evolution of Creativity and Efficiency in the Japanese Pharmaceutical Industry." *Technovation* 22: 607–14.

Takeuchi, H. and Nonaka, I. (1986). "The New Product Development Game." *Harvard Business Review* 64: 137–46.

Thomke, S. H. (1997). "The Role of Flexibility in the Development of New Products: An Empirical Study." *Research Policy* 26(1): 105–19.

Varela, J. and Benito, L. (2005). "New Product Development Process in Spanish Firms: Typology, Antecedents and Technical/Marketing Activities." *Technovation* 25: 395–405.

Walsh, J. P. and Dewar, R. D. (1987). "Formalization and the Organizational Life Cycle." *Journal of Management Studies* 24(3): 215–31.

Walsh, S. T. and Linton, J. D. (2001). "The Competence Pyramid: A Framework for Identifying, and Analyzing Firm and Industry Competence." *Technology Analysis & Strategic Management* 13(2): 165–77.

Wheelwright, S. C. and Clark, K. B. (1992). "Creating Project Plans to Focus Product Development." *Harvard Business Review* 70(2): 70–82.

Yam, R. C. M., Guan, J. C., Pun, K. F., and Tang, E. P. Y. (2004). "An Audit of Technological Innovation Capabilities in Chinese Firms: Some Empirical Findings in Beijing, China." *Research Policy* 33(8): 1123–40.

Yin, R. K. (1994). *Case Study Research: Design and Methods* (2nd edition). Thousand Oaks, CA: Sage Publications.

Yin, R. K. (2003). *Case Study Research: Design and Methods* (3rd edition). Thousand Oaks, CA: Sage Publications.

7

Roadmapping at Printco: One Company's Experience

Rick Mitchell, Robert Phaal, Clare J. Farrukh, and David Probert

Overview

Printco develops and manufactures printing solutions for industrial applications, operating in more than 150 countries. The company headquarters are in Europe, colocated with core design and manufacturing operations, with regional centers and sales and support organizations based around the world. Printco employs 1500 staff, with a turnover in 2006 of more than £200M. The company is 30 years old, and has a strong technology heritage. As the company has grown in size and complexity, new technologies have been acquired and the product range expanded, with a need to establish methods to manage the effective acquisition and integration of technology into the core new product introduction process.

This chapter describes how technological considerations are crucial to the ongoing development of innovative and successful products in the company, focusing on a recent technology acquisition associated with the creation of a new business unit. This highlighted the need for an effective business planning approach to ensure integrating technology and product development. The implementation of roadmapping in the organization over a period of five years is presented, summarizing key learning points.

The search for new technologies

In the early 1990s Printco's management started to be concerned about the restrictions that their core technology would place on the growth of the company. Their core technology was a flexible "continuous ink jet" printing process, ideal for printing simple information such as date codes onto products moving on a production line. But it had serious limits: the resolution was relatively low compared with most printing methods and it was limited to relatively small characters. At that time there were also other disadvantages such as reliability and some environmental concerns. There was a clear demand in the market for larger characters as well as images such as bar codes and logos, which could not be met by any existing flexible printing method.

The board recognized that the fundamental limits on the performance of their core technology posed both a threat and an opportunity for the company. Without a new technology the company's rapid growth could soon come to an end. But there was obviously a demand for better performance if only it could be achieved. The big danger was that if a competitor moved in with something better, Printco would not only miss out on a new business opportunity but could lose many of its existing customers.

During the 1990s Printco's technical staff looked for ways to print larger, higher-definition images onto moving products. One option was to build a more sophisticated (and much more expensive) version of the core product that would give great printing speed and could be expanded to large printing widths. The equipment would be much more complex than their existing products but there would be good applications in the commercial printing industry in which the company already had some presence.

Another possibility was to adopt a variant of the "drop on demand" technology used in desktop printers. This was a high-resolution technology and the print-heads could be stacked together to print large images. But it was slow, very sensitive to the distance from the print-head to the surface, and could print only onto paper or cardboard.

A final possibility was laser marking. This works by rapidly scanning a small spot of laser energy over the product. It makes a mark by removing a layer (for example of printed ink) and exposing the surface beneath, or by changing the color of the surface itself. This technology is fast, reliable, and environmentally friendly but is not suitable for all surfaces. One further drawback was that laser marking requires no ink, so Printco would forgo a very important source of revenue. Another was that nobody in the company knew much about lasers, either from a technical or marketing perspective.

The board soon realized that there was no simple solution to their problem. There could be no single replacement for their core technology; instead they would have to enter several new fields at once. The easiest decision was to develop the expensive and sophisticated version of their mainline printing system because it was within the technical capabilities of the in-house R&D team, and there was a clear demand for higher speed and higher resolution printing in the commercial printing market where the higher price would be acceptable. Developing this technology proved more of a challenge than expected but Printco eventually launched a product which became successful and substantially replaced the existing technology in the commercial printing sector.

Laser was clearly a case for acquisition and Printco bought a small laser company in the USA that had a unique small high-power laser tube already used in marking equipment. Printco's international distribution and knowledge of the marking market together with the new subsidiary's technical expertise allowed them to take a leading position in the expanding laser marking market.

However, it proved more difficult to find a suitable form of the drop on demand inkjet technology. Printco's technical team surveyed all the available examples in the early 1990s and found none that met their requirements. Eventually, however, suitable print-heads did come onto the market and Printco began to build them into products that could print high-resolution images onto paper and cardboard packaging.

Each of the new technology areas posed its own unique set of problems and opportunities so in the first instance each was established as a separate division to give the maximum autonomy to develop in its own way. Divisional managers had considerable discretion over distribution, product policy, and commercial strategy.

Printco's need for a technology planning process

Printco first recognized the need for an improved technology and product strategy in the late 1990s. A few years earlier a planning process had been introduced that for the first time documented short, medium, and long-term strategies for the company up to ten years ahead. This process achieved its aims of providing a vision of the future but it exposed gaps that demanded attention. Most critically, it showed that some important objectives existed in a vacuum: they were identified as important but did not necessarily have programs in place to make them happen.

As a technology-based company, Printco was particularly aware that developing new technologies – or other competences – could take a long time. The company had had experience of including new technologies in product development projects before they were fully tried and tested. The result had always been delay and disappointment. To avoid this it was clear that they needed a coherent product-technology strategy so that innovations could be developed in advance and then brought to market quickly and securely when required.

Selecting the process

The need for better product-technology planning was felt in different ways in different parts of the company. The core technology – used by the largest and most profitable part of the company – was known to be near the limit of improvement in terms of the key printing parameters: resolution, speed, and print size. A host of peripheral improvements were possible, but there was little opportunity for dramatic impact on the core functions of the product. The result was a feeling of drift. What was the right way to go, and was there in fact any viable strategy other than cost reduction?

The laser division, by contrast, had slim resources and a technology that was still improving rapidly. Their problem was to choose the right path through a multitude of opportunities.

The other two divisions, however, were very much occupied at the time with a mix of operational problems and opportunities and their management had little appetite for long-term planning. However, everyone knew that this situation would change, so any new process would have to be suitable for these divisions too.

The company needed to find a method for developing the product and technology strategies that would be suitable for all four divisions despite the fact that their technologies and markets were at very different stages of maturity. These strategies would initially be separate but there would certainly be synergies between them which might be important in the longer term.

After a period of searching, Printco's directors chose "roadmapping" as a likely candidate for their technology strategy process. The most immediately attractive feature for them was that the physical product would be quite literally at the centre of the plan. This fitted well with the way all parts of the company approached the business. Technologists, production staff, marketers, and salesmen all thought about it in different ways but the product was the shared medium of discourse for the whole company.

Technology roadmapping

Technology roadmapping, and its many derivatives, has become one of the most widely used approaches for supporting innovation and strategy, at firm, sector, and national levels. The roadmapping approach, first developed by Motorola more than 25 years ago, has been adopted (and adapted) by many organizations, initially within other large technology-intensive firms in the consumer electronics, aerospace, and defense sectors.

A key benefit of roadmapping is the communication associated with the development and dissemination of roadmaps, particularly for aligning technology and commercial perspectives, balancing market "pull" and technology "push." Roadmaps can take many forms, but the most general and flexible approach comprises a time-based, multilayered chart, illustrated in Figure 7.1, enabling the various functions and perspectives within an organization to be aligned, and providing a structured framework to address three key questions: Where do we want to go? Where are we now? and How can we get there?

The type of roadmap illustrated in Figure 7.1 emphasizes the need to take a multifunctional view, as technology without application has limited

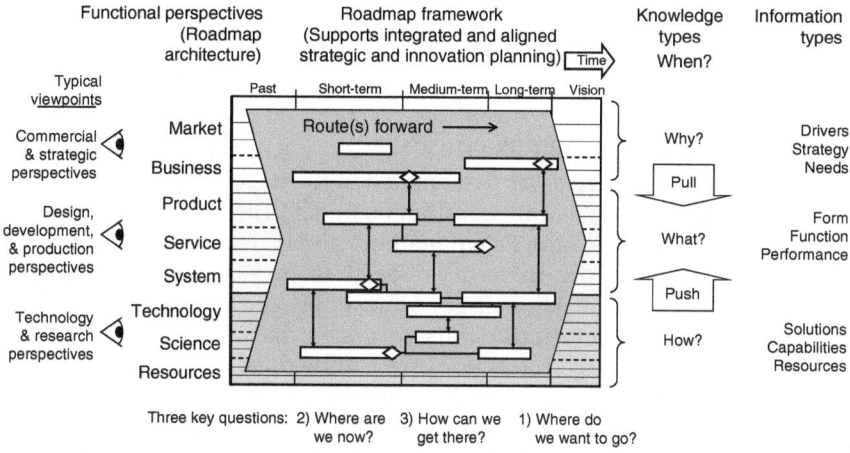

Figure 7.1　Typical multilayered time-based roadmap, linking technology to markets, providing an integrating framework to support alignment of perspectives across the organization

value. Thus, describing this as a technology roadmap is somewhat limiting, and other terms such as strategic, business or innovation roadmapping are becoming more widespread.

Workshops are an important element of any roadmapping process, providing a forum for strategic dialogue across functions and business units, using the roadmap to structure the discussion. The University of Cambridge Centre for Technology Management has been developing workshop-based approaches for supporting the initiation of roadmapping processes since 1998, in collaboration with industry. This Chapter describes the first such collaboration, where the "T-Plan fast-start" approach (Phaal et al., 2001) was first piloted, and the subsequent evolution of the technique in the firm.

The T-Plan workshop approach focuses on integrated product-technology planning, bringing together 8–12 participants from across the organization to develop a "first-cut" roadmap for a product area, in four half-day workshops:

Market: identify and prioritize external market and internal business drivers, and review business strategy.

Product: identify potential product features, functions, and attributes, and prioritize these with respect to how strongly they address the drivers.

Technology: identify and prioritize potential technological solutions for developing the product features.

Charting: based on the outputs from the first three workshops, develop a "first-cut" roadmap, linking market, product, and technology perspectives.

Rolling out the roadmapping process

The roadmapping process was first applied in the largest division of Printco, which is based on the mature continuous inkjet printing technology. This was the first full pilot of the T-Plan method, and constituted an experiment for both the firm and the university research team. The process was applied over a series of four weekly half-day workshops, and a first draft roadmap emerged without too much difficulty.

The divisional manager's reactions were immediate and surprising: "First of all, it's too complicated," she said, "there are too many projects on it and we don't have the time or resources to do them all effectively. We must concentrate on doing fewer things better. And anyway, now that I see the whole picture, I'm not sure that our assumptions about the drivers in this market are up-to-date."

She launched a concerted review of customers' requirements using cross-functional teams helped by outside market research consultants.

Product feature concepts	1. Market driver 1 (P:7)	2. Market driver 2 (P:3)	3. Market driver 3 (P:0)	4. Market driver 4 (P:2)	5. Market driver 5 (P:6)	6. Market driver 6 (P:5)	7. Market driver 7 (P:1)	8. Market driver 8 (P:8)	A. Business driver 1 (P:10)	B. Business driver 2 (P:8)	C. Business driver 3 (P:9)	D. Business driver 4 (P:5)	Market	(N)	Company	(N)
1. Feature area 1	?						✓		✓✓			✓	1	0	25	4
2. Feature area 2	✓✓						✓✓		✓	✓✓✓		✓	16	4	39	6
3. Feature area 3		✓✓✓			✓				✓	✓✓	✓		15	4	35	5
4. Feature area 4	?		✓✓✓			✓✓	✓✓	✓	✓✓✓	✓✓	X		20	5	37	6
5. Feature area 5	?					✓✓✓	✓	✓✓✓	✓	✓	?		40	10	18	3
6. Feature area 6	?	✓			✓	✓	✓✓		✓	✓✓		✓	16	4	31	5
7. Feature area 7	✓✓	✓		✓	✓	✓	✓✓		✓	✓	✓✓		32	8	36	6
8. Feature area 8	X	✓✓		✓	✓		✓		✓	✓	X		8	2	9	1
9. Feature area 9					✓✓✓	✓✓✓	✓✓		✓✓	✓✓	X		35	9	27	4
10. Feature area 10	✓✓	✓✓							✓✓✓	✓	✓✓✓		20	5	65	10
11. Feature area 11		✓	✓	✓	✓	✓✓	✓✓		✓✓	✓		✓	23	6	33	5
12. Feature area 12	X	✓✓		✓	✓		✓✓		✓✓	✓✓	X		9	2	27	4
13. Feature area 13	✓	✓✓	✓		✓✓	✓✓	✓✓		✓✓✓	✓✓✓		✓	36	9	59	9

Prioritization (P): 7 3 0 2 6 5 1 8 | 10 8 9 5 (out of 10). Market columns 1–8; Company columns A–D. Score = Σ ticks × P; ✓=1, ✓✓=2, ✓✓✓=3, X=−1; (N) = normalized.

Figure 7.2 Cross-impact matrix used in compiling the first roadmap for the laser division

This took several months, and showed that Printco had been over-emphasizing some features that were now less important to customers than they had previously been; but more importantly, some significant new needs were being overlooked. A revised roadmap was drawn up incorporating the new information, and a series of new product initiatives followed.

The laser division produced its first roadmap in two working days. This concentration of effort was necessary because some of the key staff worked in different sites in Europe and the USA, and it would have been inefficient for them to travel in and out repeatedly for shorter meetings. Figure 7.2 shows the cross-impact matrix used to relate the market drivers to the key product features. Figure 7.3 shows the roadmap itself (commercially sensitive information has been removed).

The divisional manager's initial reaction to the roadmap was the same as for first application: that the division was trying to do too many things (this pattern was later repeated in each of the other divisions). The first review of the roadmap, after 6 months, produced a simplified version (Figure 7.4) that formed the basis for several more iterations. The key product features also evolved over a few iterations before settling down to a stable pattern that remained, and was used as a starting point for roadmapping in other divisions. Interim reviews were generally done every 6 months, with more fundamental reappraisals at longer intervals.

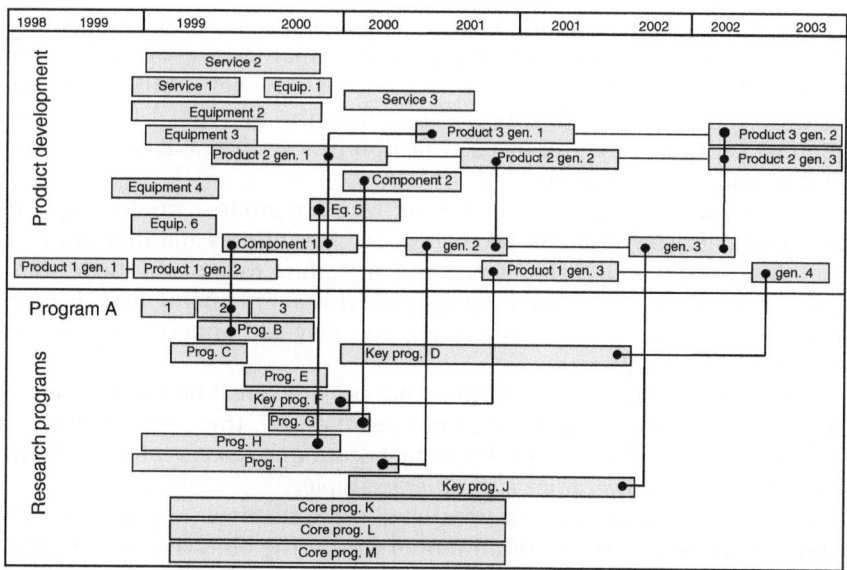

Figure 7.3 Printco's first roadmap for the laser division

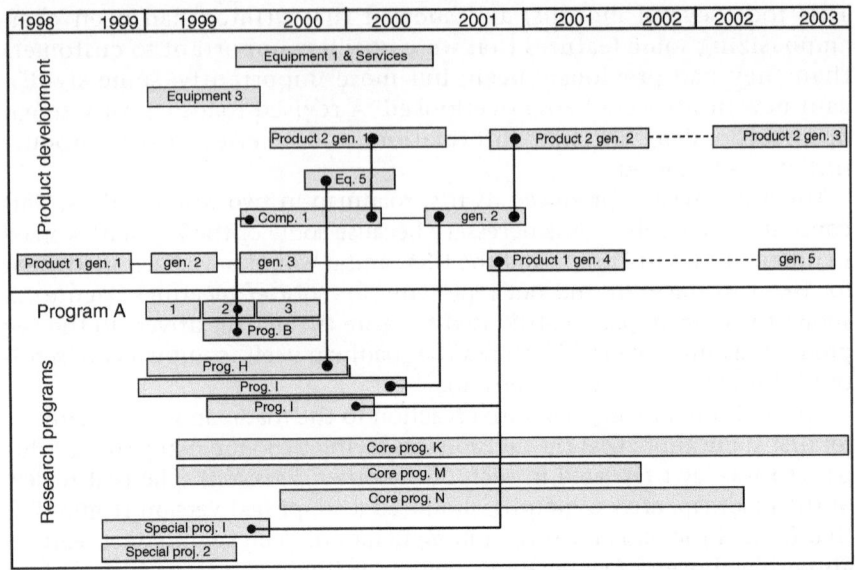

Figure 7.4   First revision of the laser division roadmap, showing the simplification and reduction of the number of projects

At the major reviews managers took particular care to plot the future for the key product features and to check whether continued improvement was likely to give competitive advantage. In this analysis they used the Kano analysis (Kano et al., 1996; see also Goffin and Mitchell, 2005) of product features (Figure 7.5). This model shows how the individual features of a product contribute to customer satisfaction (the vertical axis of the model) and fit into three categories:

Basic features: are attributes, without which a product would simply be unacceptable. The customer takes them as prerequisites and may not even mention them, though they are essential. Failure to provide them causes great dissatisfaction; however, putting effort into enhancing basic features leads to no extra customer satisfaction and may add unnecessary cost and complexity to the product.

Performance features: are features that provide a real benefit to the customer, and the more highly implemented they are, the more satisfied the customer will be. An example is the fuel economy in a car – the more efficient the better. Lower price is another example.

Excitement features: are features that give the customer unexpected value and may be attractive out of all proportion to the objective benefits they give. The problem is that traditional market research (surveys and focus groups) is unlikely to identify the requirements for excitement features.

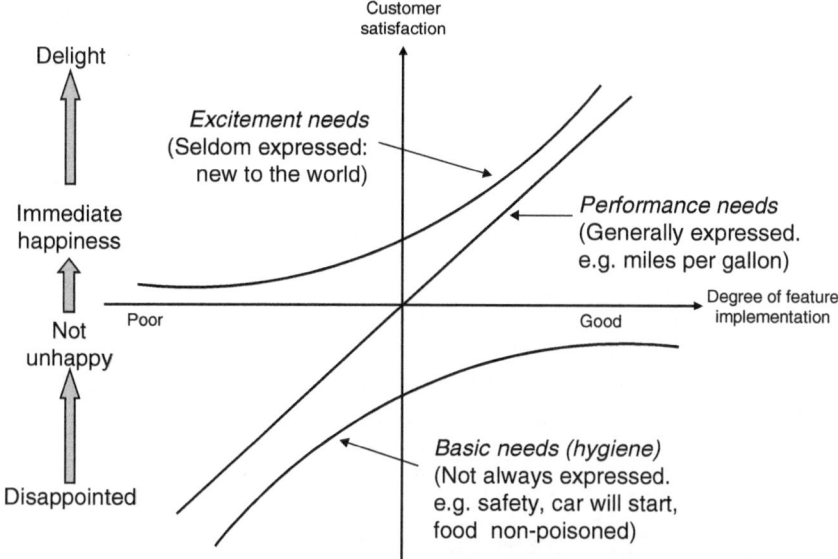

Figure 7.5 Kano's model showing contribution made to customer satisfaction by increased implementation of different types of features

It is often observed that features that start as delighters tend to become performance features as time goes on. Correspondingly, performance features often reach the point when further improvement is of little interest to customers, so they no longer provide competitive advantage. Examining their roadmaps in this light led Printco to reduce their effort on some product features and concentrate on new sources of competitive advantage.

Embedding the roadmapping process

When the roadmapping process had been in use for some time Printco purchased another company, and with it a number of product lines that were complementary, but overlapping, with their existing ones. The design of the new subsidiary's products was fundamentally different from Printco's and offered a different range of benefits. There was room in the market for both types but it would be inefficient and wasteful not to use a common electronics platform, and especially not to use Printco's proprietary printing elements as far as possible in all product lines.

Meanwhile, however, in a dynamic and competitive market both sets of products would have to continue to improve even as they were

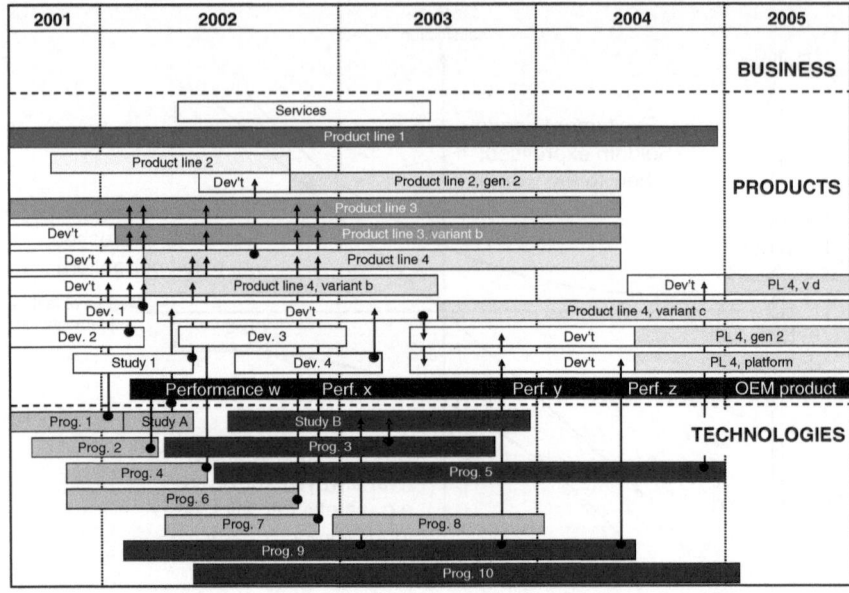

Figure 7.6 The new laser division roadmap after the merger, the different colored bars in the Product section indicate the underlying platforms – the key purpose of the new map is to bring all products to the same platform, which is finally achieved in year 5

brought together onto a common platform. This would require several steps, so a new roadmap was required that would show how the product lines could be unified without losing competitive advantage on the way. The roadmapping process proved well able to achieve this. The restructured roadmap is shown in Figure 7.6. This convergence took four years and several product updates, during which time the key performance features identified in the roadmapping process were steadily updated and improved – see Figure 7.7 (commercially sensitive detail has been removed).

The T-Plan "fast-start" process was used in all Printco divisions and proved an effective way to create initial roadmaps efficiently and quickly. In all cases the first roadmap showed that the existing plans and intentions were too ambitious and had to be scaled back – a valuable early result and a useful benefit from the work.

Nevertheless, managers and staff usually treated the first versions of the roadmaps with caution and only really trusted them after they had been through several iterations. These reviews, typically every six months, were

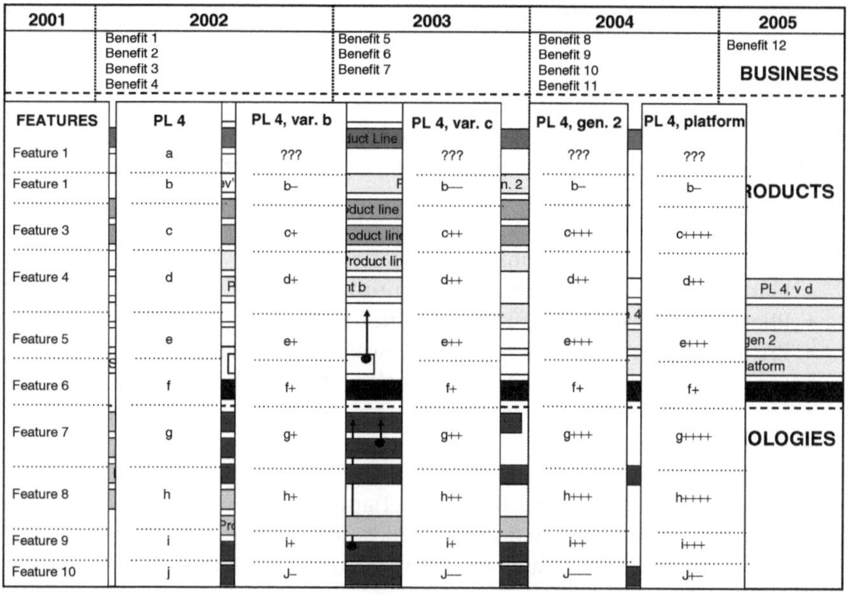

2001	2002		2003	2004	2005
	Benefit 1 Benefit 2 Benefit 3 Benefit 4		Benefit 5 Benefit 6 Benefit 7	Benefit 8 Benefit 9 Benefit 10 Benefit 11	Benefit 12 **BUSINESS**
FEATURES	**PL 4**	**PL 4, var. b**	**PL 4, var. c**	**PL 4, gen. 2**	**PL 4, platform**
Feature 1	a	???	???	???	???	
Feature 1	b	b–	b—	b–	b–	RODUCTS
Feature 3	c	c+	c++	c+++	c++++	
Feature 4	d	d+	d++	d++	d++	PL 4, v d
Feature 5	e	e+	e++	e+++	e+++	gen 2
Feature 6	f	f+	f+	f+	f+	atform
Feature 7	g	g+	g++	g+++	g++++	OLOGIES
Feature 8	h	h+	h++	h+++	h++++	
Feature 9	i	i+	i+	i++	i++	
Feature 10	j	J–	J—	J—	J+–	

Figure 7.7 Stages of the planned convergence of the product platforms, showing steady development of the key features, and associated business / market outcomes (the level of increase (+) or decrease (-) in function and performance in each feature as the products evolve is shown schematically, relative to product line (PL) 4)

crucial. They gave time for participants to gather extra data and to reflect on what had been done. Inevitably the maps evolved and stabilized with repeated discussion but the process of debate also cemented understanding and support. The roadmaps became a useful and valued tool for communicating the emerging strategy to the board of the company and to others in the division. The next stage for the company will be to bring the divisional roadmaps together to expose the synergies between them that can lead to further efficiencies.

Managerial implications

- Roadmapping is a powerful and flexible technique that is being increasingly adopted in industry as a core integrating mechanism for strategic planning and innovation.
- A principle benefit is the communication that is engendered, both during the development of the roadmap and afterwards, using roadmaps as a common reference point and language to support the ongoing dialogue

that is essential for effective innovation and strategy development and implementation.

- However, roadmapping can be challenging, due to a combination of the broad scope and complexity of the issues being addressed, uncertainties associated with the future, and gaps in available knowledge.
- The following success factors should be considered when embarking on a roadmapping initiative:
  - Establish a clear business need
  - Ensure commitment from senior management
  - Plan carefully and customize the approach to suit the circumstances
  - Phase the process to ensure that benefits are delivered early
  - Ensure that the right people and functions are involved
  - Link the roadmapping activity to other business processes and tools
  - Provide adequate support and resources
  - Keep it simple
  - Iterate and learn from experience
- The company case described in this chapter reinforces these key success factors, providing a detailed account of how roadmapping was introduced in one company, and how the process evolved over a number of years.

References

Goffin, K. and Mitchell, R. (2005). *Innovation Management: Strategy and Implementation Using the Pentathlon Framework*. Houndmills: Palgrave Macmillan.

Kano, N., Saraku, N., Takahashi, F. and Tsuji, S. (1996). "Attractive Quality and Must-be Quality." In J. Hromi (Ed.), *The Best on Quality* 7(10). Milwaukee: ASQC: 165–86.

Phaal, R., Farrukh, C. J. P., and Probert, D. R. (2001). *T-Plan: The Fast-start to Technology Roadmapping – Planning Your Route to Success*. Cambridge: Institute for Manufacturing, University of Cambridge.

8
Performance Measurement in Supply Chain Collaboration

Dilek Cetindamar, Bülent Çatay, and Osman Serdar Basmaci

Introduction

Collaboration is an advantageous strategy for technology-based competition. The management of collaboration involves the mutual efforts of supply chain partners to design, implement, and manage seamless value-added processes (Johnson and Johnson 1989). The development and integration of people and technological resources as well as the coordinated management of materials, information, and financial flows underlie successful supply chain collaboration (SCC). As the competition among firms is shifting from company versus company to supply chain versus supply chain, managers should not only adopt SCC whenever needed but also manage it successfully (Gomes-Casseres, 1994; Pagh and Cooper, 1998; Salcedo and Grackin, 2000). This necessitates, among others, a measurement system that is useful for the management of decision-making in various issues such as planning, control, alignment, and improvement of collaboration (Landeros et al., 1995; van Hoek, 1998).

Even though collaborations or alliances are widely applied, they are not studied extensively at the empirical level (Cahill et al., 1998). Studies about industrial networks, production clusters or complexes are numerous but they are, by and large, at the regional or industry level (Cooke and Morgan, 1993; Saxenian, 1994; Staber, 1996). Supply chain studies concentrating on the performance measurements in inter-firm relationships are scarce in the literature (Gulati, 1998; van Hoek, 1998). Mechanisms such as value co-production, collaborative/extended enterprises, virtual enterprises are addressed (Childe, 1998; Ramirez, 1999; Bititci et al., 2005); however the literature particularly lacks in innovation driven collaborations where technologies are developed and diffused through collaboration (Benfratello and Sembenelli, 2002; Carlsson et al., 2002; Chang, 2003). Moreover, the majority of studies are in advanced countries, ignoring the practice of inter-firm collaborations in developing countries (Caloghirou et al., 2003). If the collaborative form of management is advantageous for competition, it is of

137

high concern to understand mechanisms behind the success of collaborations. By studying collaboration experiences, it might become possible to develop and improve its management both in developed and developing countries. This paper is motivated by an interesting collaboration model in the Turkish textile dyeing and finishing industry. The collaboration under investigation is a joint venture aiming at developing and diffusing new technologies. This study presents an empirical study addressing the dynamics of this collaborative effort and develops performance metrics to measure the performance of the strategic partnership.

The remainder of the paper is organized as follows: The next section introduces the theoretical background by reviewing the literature about collaborations with a particular emphasis on their performance measurement. The third section presents the empirical study and the forth section introduces the results. Finally, the findings of the study are discussed in the conclusions section.

The performance measurement of a technological joint venture

Supply chain oriented measures should facilitate greater inter-firm collaboration. However, most firms are experiencing difficulty in devising and implementing supply chain measures (Lapide, 1998; van Hoek, 1998). Measures need to provide an accurate picture of supply chain performance as a whole but also highlight opportunities for improvement at both the individual firm and the overall supply chain levels (Lapide, 1998). Considering the inherent complexity of the supply chain, selecting appropriate performance measures for SCC is critical.

Supply chain models have predominantly utilized cost and financial measures as performance metrics while some studies also indicate a combination of cost and customer responsiveness such as flexibility and quality (Neely et al., 1995; Beamon, 1999; Tan, 2002). In some cases, empirical studies examine the termination of an alliance to measure the performance of alliances (Gulati, 1998). Due to increased importance of technologies in competition, recent literature attempts to widen performance measures by suggesting new metrics such as new product development, knowledge transfer, patents, and productivity (Womack et al., 1990; Hagedoorn and Schakenraad, 1994; Reuer and Miller, 1997; Das, 2000; Kotabe et al., 2003; Chung and Kim, 2003; Kale et al., 2002). However, there is no widely accepted set of measures in the literature, since technological partnerships are particularly difficult in defining and measuring performance metrics due to the uncertain and complex nature of knowledge and innovation developed in collaborations (Hagedoorn and Cloodt, 2003).

In the manufacturing literature it is often argued that performance measures should be derived considering strategic objectives; that is, they should be used to reinforce the importance of certain strategic variables (Neely et al., 1995; Freeman and Soete, 1997). Further, given the multifaceted objectives of many alliances, performance can be difficult to measure with financial outcomes alone (Hagedoorn and Schakenraad, 1994; Benfratello and Sembenelli, 2002). That is why following the model of a recent study that develops measures on the basis of strategy of alliance and the operational contribution of players in the supply chain competitiveness (van Hoek, 1998), this study will focus mainly on strategic objectives in deriving performance measures for a technology collaboration.

In addition, measures are in general used as individual performance criteria, while few studies combine them in a framework. These frameworks, such as the balanced scorecard (Kaplan and Norton, 1992) and the performance prism (Kennerley and Neely, 2000) have been proposed with the objective helping organizations to define a set of measures that reflects their objectives and assesses their performance appropriately. The frameworks are multidimensional, explicitly balancing financial and non-financial measures. Following this system idea, as shown in Figure 8.1, this study proposes a model where the performance measures for an alliance includes metrics for (1) internal performance, (2) perceptions of partners regarding the alliance performance, and (3) the degree of inter-firm relationship.

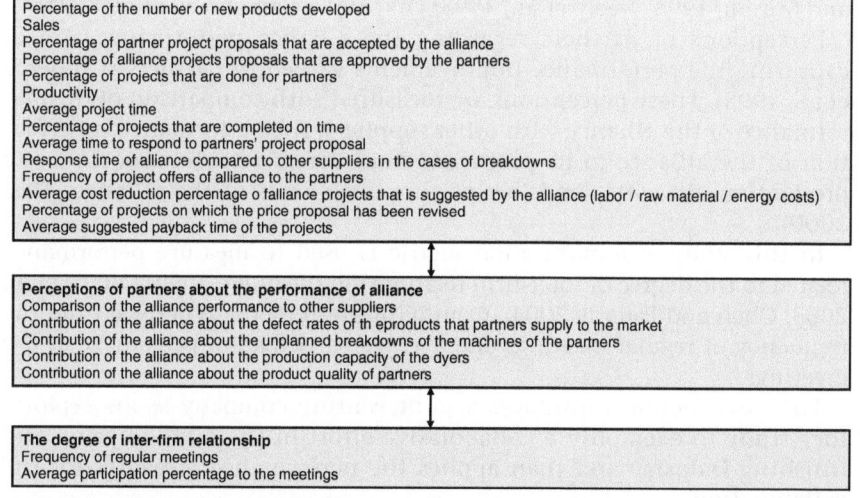

Internal performance
Innovation
Percentage of the number of new products developed
Sales
Percentage of partner project proposals that are accepted by the alliance
Percentage of alliance projects proposals that are approved by the partners
Percentage of projects that are done for partners
Productivity
Average project time
Percentage of projects that are completed on time
Average time to respond to partners' project proposal
Response time of alliance compared to other suppliers in the cases of breakdowns
Frequency of project offers of alliance to the partners
Average cost reduction percentage o falliance projects that is suggested by the alliance (labor / raw material / energy costs)
Percentage of projects on which the price proposal has been revised
Average suggested payback time of the projects

Perceptions of partners about the performance of alliance
Comparison of the alliance performance to other suppliers
Contribution of the alliance about the defect rates of th eproducts that partners supply to the market
Contribution of the alliance about the unplanned breakdowns of the machines of the partners
Contribution of the alliance about the production capacity of the dyers
Contribution of the alliance about the product quality of partners

The degree of inter-firm relationship
Frequency of regular meetings
Average participation percentage to the meetings

Figure 8.1 Performance measurement system for a SCC in technology development

While internal performance might be considered the static part of the performance measurement system, second and third measures of the model brings dynamic evaluation into the system as suggested in the literature (Bititci et al., 2000; Kennerly and Neely, 2002). It becomes a need to have a performance measurement system that is dynamic enough to reflect changes in the internal and external environment as well as helps to review and prioritize objectives as the environment changes.

In the model, internal performance is measured with productivity, innovation, and sales (Neely et al., 1995; Beamon, 1999; Chung and Kim, 2003; Kotabe et al., 2003). Productivity is measured with a number of indicators related to product quality, project time, and so on as shown in Figure 8.1. As the goal is to measure the performance of a collaboration to develop high value-added products and technologies, innovation performance becomes a core measure and it is measured with percentage of the number of new products developed (Das, 2000; Gunasekaran et al., 2001). Sales performance might be measured with sales or purchasing volume generated through partnership.

As mentioned in the literature, partners might have a different performance perception about the alliance (Dollinger and Golden, 1992; Medcof, 1997). Different members of the supply chain team consider performance of alliance on the basis of their own individual perceptions. The typical result is channel conflict and perhaps even the dissolution of the relationship. Thus, it is important to know what is truly valued and then put the right measure in place (Lambert and Cooper, 2000). Without disregarding the objectives of individual partner firms and the alliance itself performance metrics should be used for mutual benefits (von Hippel, 1988; Heide and Stump, 1995; Neely et al., 1995; Olk, 1997; Caloghirou et al., 2003).

Perceptions of partners regarding the alliance performance aim at capturing the performance improvements gained in partners (Landeros et al., 1995). These perceptions are measured with comparison of the performance of the alliance with other suppliers, the performance contribution of the alliance to its partners (about defect rate, breakdowns, and production capacity), and low performance areas of the alliance (Das, 2000).

In this study, communication metric is used to measure performance related to the degree of inter-firm relationship (Beamon, 1999; Kotabe et al, 2003; Chen and Paulraj, 2004). Communication metric is derived from the frequency of regular meetings and average participation percentage to the meetings.

The next section introduces a joint venture company as an exploratory study to exemplify a collaborative effort in the textile dyeing and finishing industry and then applies the performance measures for the collaboration.

The empirical study

The textile and apparel sector has been the backbone of the Turkish economy with a vital role to play in the industrialization process and market orientation of the economy in the last two decades. The textile sector continued to be one of the major contributors to the Turkish economy, being one of the fastest growing sectors in the 1990s with an average 12.2 percent annual growth, while the Turkish economy had an average growth of 5.2 percent per year (Foreign Economic Relations Board, 2002). The sector has great significance in terms of economic development with a share of GNP above 10 percent, industrial production around 40 percent, manufacturing labor force around 30 percent and exports around 35 percent.

The advantages of Turkish textile industry are numerous like its geographic position, low labor costs, growth of high quality cotton, and so on but those advantages are insufficient in this highly competitive environment. In addition to these advantages, Turkey must also acquire the competitive strengths like high quality, high technology, quick service, dynamic structure, production flexibility, high productivity, high R&D accumulation and innovative product development talents (Foreign Economic Relations Board, 2002).

Here comes the importance of our exploratory study firm: Textile Dying Technologies (3T). Four companies (namely, Pisa Textile, Ekoten Dyeing, Eliar Electronics, and Vega Machinery) felt responsible to undertake the mission of improving the Turkish dyeing and finishing industry and decided to make an industry-wide collaboration by founding a joint venture at the end of 2000.

In 2003, 3T had six technology supplying and ten dyeing and finishing partner companies, each having one-sixteenth of a share of 3T. Technology firms are suppliers to dyeing and finishing firms that represent buyers in a typical supply chain. This is interesting regarding 3T because of two reasons. First, it is more than a buyer-supplier relationship since it involves collaborative development of technologies. Second, 3T is formed by large and small companies together contrary to many examples seen in the literature where alliances are formed among small firms (Malecki and Tootle, 1996) or among one large company and many small firms as the case of Toyota supplier network (Dyer, 2000).

Collaboration has become an established aspect of innovation strategy (von Hippel, 1988; Harris et al., 2000; Caloghirou et al., 2003). Literature identifies that technological complementarity of partners, concrete development of innovations, and the need for technology monitoring are important motives for forming strategic alliances (Hagedoorn and Schakenraad, 1994; Dyer and Singh, 1998). This is also observed in our example. The first aim of 3T is to help the enterprises increase the productivity and efficiency. Integrated automation systems are supplied to the textile dyers by 3T. It conducts feasibility studies and offers after sales services as well. Starting out from the idea that the textile dyers and finishers have to produce innovative and high-tech products, it is concluded that the systems which allow production flexibility, high productivity and efficiency, high technology, and dynamic structure must be implemented at dyeing and finishing facilities in Turkey to remain competitive in global markets. So, the mission of 3T is to help the dyers, which possess this vision, with all their efforts and opportunities. It develops and implements automation solutions not only to its ten partner dyers, but also other dyers in Turkey by utilizing the technology of its six supplier partners as well as other suppliers, when needed.

This also enables 3T to expose a collaboration model that is built on technology development, technology intelligence, and technology diffusion among firms in a supply chain (von Hippel, 1988; Chung and Kim, 2003). In order to realize this sharing and diffusion, 3T meets with its partners regularly. In these meetings, partner dyers share customer needs and fashion trends they have determined with other partners and partner suppliers share their technological innovations and capabilities. Then, 3T matches these customer needs and fashion trends with the supplier capabilities and develops new products and/or solutions by the help of suppliers. First, 3T uses the products of its partner suppliers. However, it also uses additional foreign and domestic resources for the products that it cannot supply from its partner suppliers.

In early 2003, several semi-structured interviews were conducted with the managers of 3T, Pisa Textile (dyer), and Eliar Electronics (supplier) to obtain some insights about the collaborative model and to understand its mechanisms. Then, three questionnaires were prepared based on the information gathered from these interviews and the literature review. The first was for 3T, as the collaboration unit; the second was for the partner and non-partner dyers; and the third one was for the partner suppliers. After three questionnaires were conducted during the period of May and July 2003, a set of interviews was conducted again with the managers of 3T, Pisa Textile, and Eliar Electronics to discuss the results. The results were also presented in a board meeting to all partners of 3T to receive their comments.

The questionnaire for dyers was conducted to 70 dyers and they were sent and received via fax and e-mail. Out of 70 dyers, only ten are the partners of 3T, the remaining ones are customers of 3T. All questionnaires were originally addressed to the general manager of the company or to the director of manufacturing and were mostly completed and returned by a contact person holding a managerial position in the manufacturing area. To increase the response rate, follow-up phone calls were also made. In total

30 responses (of which ten are partners of 3T) were received from the dyers with a response rate of 42.8 percent.

The questionnaire for suppliers was prepared for six partner suppliers of 3T. These questionnaires were also sent and received via fax and e-mail. The response rate of this questionnaire is 100 percent.

The total respondents for the questions regarding performance measures were 16 companies, all partners of 3T, while 36 companies altogether answered questions regarding collaboration in general (Tables 8.3 through 8.5 in Section 8.4).

The results of the empirical study

General data

Our data show that the six partner suppliers are mainly small firms employing 7 to 68 employees. Half of (52 percent) all dyeing firms are small and medium sized enterprises with less than 250 employees, while the rest are large companies with 9 percent of them employing more than 1000 employees.

We observe that the sales figures of the dyers are relatively low compared to the employment levels. 65 percent of the dyers have total sales of less than 20 million USD in 2002 and only 15 percent achieved more than 60 million USD in sales. The reason might be the fact that the sector is trying to recover the economic crisis of 2001 in Turkey. A similar reasoning may be claimed for the low level of R&D expenditures. The dyers spent on average 2.1 percent of their sales to R&D in 2002. On the other hand, a significant difference between the partners of 3T and other respondents to the questionnaire is observed in the ratio of the R&D budget to total sales: while other companies (20 dyers) allocate more budget to R&D, the partner companies (10 dyers) have transferred their R&D works to 3T and its suppliers.

Figure 8.2 depicts the current level of collaboration of the 30 dyers that responded the questionnaire. While 32 percent of the companies collaborate with more than five companies, 56 percent do not practice any sort of collaboration at all (that is, only three out of 20 non-partner dyers claim collaboration). An important issue that the partners and other companies differ from each other is the number of companies they collaborate with. Collaboration is a new concept for the textile industry as it also is for other industries in Turkey. Collaboration requires a high level of trust and information sharing, but it is not very easy to build such a trustworthy atmosphere in this business environment. So, companies are reluctant to collaborate. Since 3T has 16 partners, it is not very surprising that there is a significant difference about this variable between the partners and other companies. 68.4 percent of the other companies do not collaborate with any company; however 83.3 percent of the partners collaborate with more

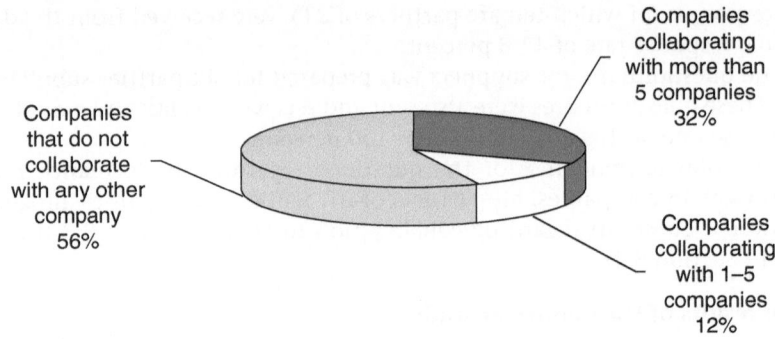

Figure 8.2   Collaboration level of dyers and suppliers

than five partners. This may certainly be due to their partnership to 3T, but the most interesting fact is that 16.7 percent of the partners stated that they collaborate with between one and five companies, although 3T has 16 partners. This means that they do not see all of the partners of 3T as their partners and it may be a significant problem for the future of this collaboration model.

Table 8.1 shows that the bid evaluation, which is the strategy where the bids are evaluated only for that single purchasing, has been the most widely used strategy among the dyers in the past. However, the technological capabilities of the suppliers and strategic collaboration appear to gain importance in supplier selection. It may be claimed that the importance of collaboration is well understood by 3T dyer partners as 90 percent stated that they would seek collaborative strategies in the next two years. Thus, collaboration is a new but rapidly developing concept for the dyeing and finishing industry in Turkey.

Even though collaboration is a new approach for the textile industry, dyers and suppliers formed 3T. When the partners are asked why they have decided to collaborate, we observe in Table 8.2 that the "necessity to compete with the foreign textile industries" is the most important reason for collaboration. Since the competition in the textile industry will increase in the world after the removal of the quotas in year 2005, companies have to increase their productivity and produce high value-added products to survive. The partners of 3T seem to be aware of this fact, since 75 percent of them see "producing high value-added products" as an extremely important production target, while other companies do not see this target as important as the partners do.

This result is in fact very significant considering the fact that the foreign trade quotas will be removed at the beginning of 2005. Therefore, Turkish dyeing and finishing industry has to reach a level at which it can compete

Table 8.1 Supplier selection strategies of the partner dyers

Supplier selection strategy	Last 2 years	Next 2 years
Bid evaluation	45%	20%
Technological capability	27%	60%
Common value creation	18%	20%
Strategic collaboration	36%	90%

Table 8.2 The mean, standard deviation, and the percentage distribution of partners' ratings of the reasons for collaboration

Reasons	Mean	Std. dev.	(%)*				
			1	2	3	4	5
Necessity to compete with the foreign textile industries	4.20	0.94		7	13	33	47
Willingness to increase supply chain and company productivity	4.13	1.25		20	7	13	60
Existence of a beneficial collaboration opportunity	4.07	0.80			27	40	33

* The percentages may not add up to 100 because of rounding.

with the foreign textile industries. The road to this level certainly passes through the collaboration that leads companies to high productivity, high quality, high innovation, and responsiveness. Thus, collaborative networks such as 3T are very crucial and necessary for Turkish textile industry as well as for other industries.

Performance of 3T

As mentioned in Figure 8.1, the performance results of 3T is grouped under three categories (1) internal performance, (2) perceptions of partners regarding the alliance performance, and (3) the degree of inter-firm relationship.

Internal performance

3T aims at increasing the productivity and efficiency of the dyeing and finishing companies through integrated automation systems to produce innovative and high-tech products. In the past two years, 3T has facilitated the development of six new products in 17 projects. Although it is a

Table 8.3　Evaluation of 3T regarding innovation, sales, and productivity performance measures

Innovation	
Percentage of new products developments among all projects conducted (in 2 years)	35%
Sales	
Percentage of dyer project proposals that are accepted by 3T	100%
Percentage of 3T projects proposals that are approved by the dyers (in 2 years)	11% (by 15 dyers)
Percentage of projects conducted with partner dyers (in USD)	43%
Productivity	
Average project time	6 months
Percentage of projects that are completed on time	75%
Percent delay in the duration estimate)	20%
Average response time to dyers' project proposal	2 weeks
Response time of 3T compared to other suppliers in case of breakdowns	Same
Frequency of project offers to the dyers (in 2 years)	158 projects (to 70 dyers)
Average cost reductions	
Labor costs	30%
Raw material	10%
Energy costs	20%
Average estimated payback time	12 months
Number of projects for which price proposal has been revised	None

short period of time to observe performance results, 35 percent new product development performance is well in accordance with the company's objectives.

In Table 8.3, we observe that 3T has never refused any proposal from the dyers in the past two years and it responds to the dyers' project proposals in two weeks on average. On the other hand, only 11 percent of the 3T project proposals (158 proposals in total) have been approved by the corresponding dyers. This shows that 3T has been extensively marketing its services and solutions in an attempt to create awareness in the industry. It appears that it has been successful in these attempts since more than half of its project volume (in USD) has been with non-partner dyers (57 percent). This also shows that the partners are more reluctant in doing projects with 3T compared to other dyers.

The average duration of the 3T's projects is six months. 25 percent of the projects have been delayed and the delay is about 20 percent of the estimated duration. Considering the fact that 3T is a new company with a different collaboration model in the industry, this is an acceptable flaw in the performance. The price offer has not been revised in any of the project proposals. This shows that 3T offers realistic and reasonably low prices to the dyers.

We have observed that 3T makes a cost/benefit and payback analysis of the project and presents it to any potential customers in their project proposals. When we analyze the projects proposed in the past two years, we observe that 3T estimated reductions of 30 percent, 10 percent, and 20 percent in labor, raw material, and energy costs, respectively. We do not have data regarding the exact reductions in these cost items, but Figure 8.3 depicts a comparison of the realization levels to the estimations of 3T based on the perceptions of partners. The most accurate estimations are seen in the labor costs (in nine out of 11) followed by the energy costs (in seven out of 11). Regarding the raw material and energy costs, actual reduction is never realized more than the estimate. We also observe that in seven out of 11 projects 3T overestimated the reduction in raw material costs. The accuracy of 3T's projections is an important factor for its reliability and credibility for future projects. Therefore, keeping in mind there is also a learning curve effect, 3T should place more care in its cost estimations.

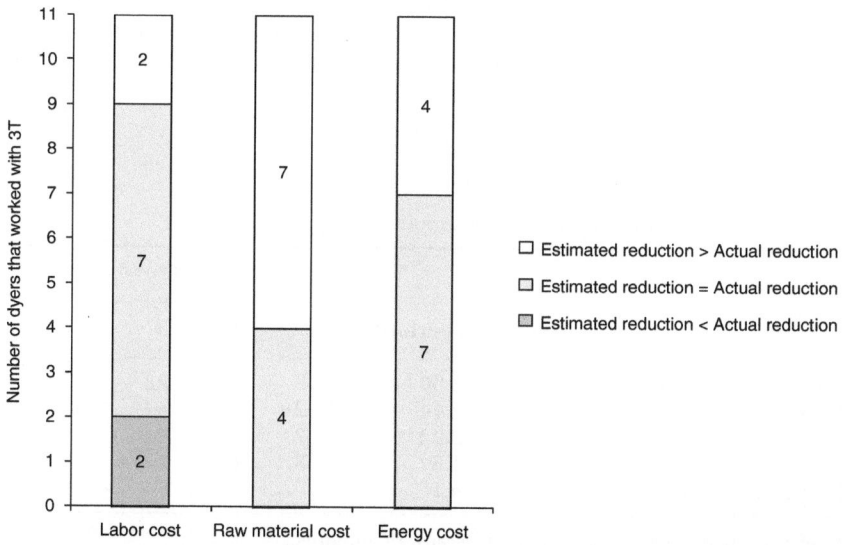

Figure 8.3 Realization level of the cost reductions in labor, raw material, and energy costs

Perceptions of partners regarding the performance of 3T

We used a 5-point Likert scale in order to compare 3T's performance in terms of technical experience, price, quality, service, and financial support (credit terms, advance payment amount, and so on) to other suppliers of the partner dyers. As observed in Table 8.4, the only criterion other suppliers outperform 3T is "financial support." In terms of "technical experience" and "service," 77 percent of the partner dyers think that 3T is better/much better than other suppliers. Also, all of the partner dyers consider other suppliers no better than 3T in "technical experience," "quality," and "service" criteria.

In what follows, we will discuss the partners' perception of the 3T's contribution to partner firms' defect rates, breakdowns, production capacity, and product quality. Figure 8.4 provides the data to evaluate the performance contribution of 3T to its partners. Five of 11 dyers stated that their defect rate of their products decreased up to 5 percent after 3T projects and the remaining six did not see any contribution. The reason behind this relatively low performance might be the different scopes of different projects since the goals of implementing automation systems may differ from one dyer to another. Regarding the contribution of 3T projects to the breakdowns of the machines and automation systems of the dyers, six out of 11 dyers stated that the percentage of breakdown duration of the machines in total production time reduced after the implementation of automation system supplied by 3T. Three of 11 projects decreased the breakdown percentages by more than 5 percent. This result is not very surprising since 3T supplies automation systems to the dyers. Regarding the production capacities, four of the dyers realized an increase of more than 10 percent while two realized an increase up to 10 percent after 3T projects.

Table 8.4   Comparison of 3T to other suppliers

Criterion	Average superiority	(%)*				
		1	2	3	4	5
Technical experience	2.00	22	55	22		
Price	2.44	22	22	44	11	
Quality	2.22	22	33	44		
Service	1.89	33	44	22		
Financial support	4.11				89	11

* The percentages may not add up to 100 because of rounding.

Note: The scale is "1=3T is much better," "2=3T is better," "3=equivalent," "4=other suppliers are better," and "5=other suppliers are much better."

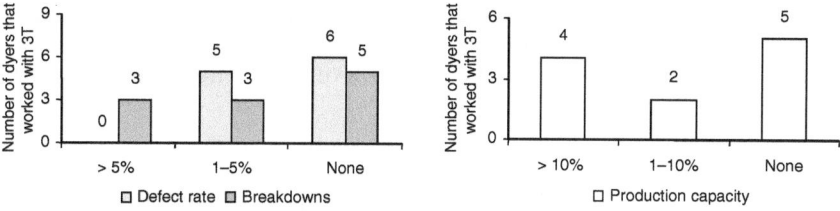

Figure 8.4 Performance contribution of 3T projects to the dyers regarding the decrease in defect rate and breakdowns and increase in production capacity

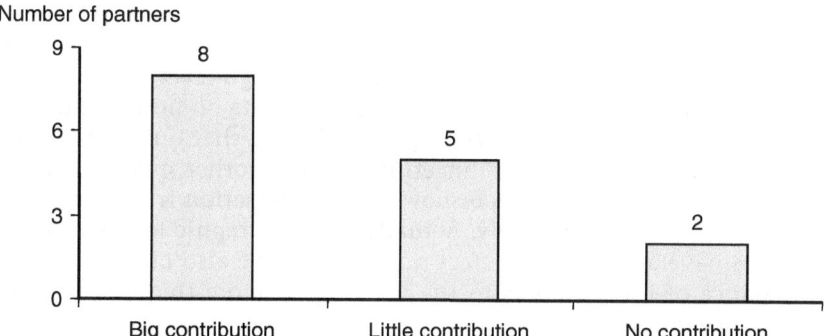

Figure 8.5   Performance contribution of 3T to its partners on product quality

In evaluating the performance contribution of 3T, we used a 5-point Likert scale. "Launch of high quality products and services" and "increase in productivity" are the top performing contribution areas as perceived by 66.7 percent and 58.3 percent of the partners, respectively, to achieve big/very big contribution. This is not a surprising result since 3T aims at letting its partners to produce high quality and value-added products through technology sharing and increasing their productivity. However, these percentages seem to be rather low. The reason might be the fact that 3T has completed a small number of projects so far.

Regarding the quality of the products 13 out of 15 respondent partners stated that they have experienced improved quality in their products after having initiated collaboration with 3T (Figure 8.5). This result has great significance considering the goals of 3T. Further, there is a strong correlation (coefficient is 0.813) between the firms' performance target "increasing the conformance to quality standards" and 3T's performance improvement area "launching of high quality products and services." This shows that the companies aiming to increase the quality level reach their targets by being a partner of 3T.

The degree of inter-firm relationship

3T organizes regular bimonthly meetings with its supplier and dyer partners. These meetings are important for the success of the collaboration since partner dyers share information about customer requirements while partner suppliers share their technological innovations in these meetings. Technology and information sharing is the main infrastructure of this collaboration model. Thus, these meetings play a significant role in facilitating partners building trust on each other and increasing their understandings about the collaboration concept. However, while 3T attends to all meetings, partners do not show such a high performance, which appears to be a barrier in building high-level trust and sharing information among partners. We collected data on how frequent the partners participate to the meetings. We observe that only a quarter of the partners "participate all" meetings while another quarter "participate most of them." Figure 8.6 shows the participation frequency of the dyers and suppliers separately. Actually, the geographic locations of the partners have a significant effect on this fact. We also observe that the attendance of the suppliers to the meetings is more frequent than the attendance of dyers.

Participants that give high importance to "willingness to share information with partners" also participate to most of the regular meetings of 3T. These companies are able to comprehend the importance of the meetings for information sharing and building trust among the partners.

When the frequency of the participation in the meetings increases, the level of the trust and information sharing between companies increases.

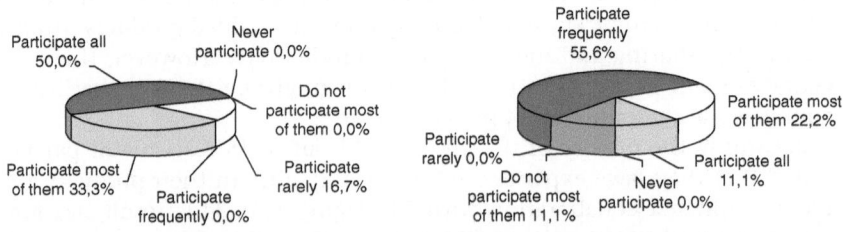

Figure 8.6   Participation frequencies of the partners to 3T meetings

While the coefficient of correlation between participation frequency to the meetings and trust level among partners is 0.982, it is 0.945 between participation frequency to the meetings and information sharing level.

In order to understand the effect of trust, information sharing, and communication on the performance of collaboration, an additional variable is used to keep track of the average of the answers of each participant to the question about performance contribution of 3T. Then, the correlation analysis was executed between this variable and variables about level of trust, information sharing, and communication. The coefficient of correlation between the statement "there exists a high level of trust between the partners" and average performance contribution of 3T is 0.810. This shows that as trust among partners increases, performance of the collaboration also increases. The correlation between the statement "we share the information regarding the requirements of the customers with our partners" and the same additional variable is also very strong with a coefficient of 0.989. Lastly, according to the analysis, the correlation coefficient is 0.887 for the average performance contribution of 3T and the participation frequency of the partners to the regular meetings. Consequently, all these results show that as the trust, information sharing, and the communication between the partners increase, the performance contribution of collaboration to the partners also increases (Sako and Helper, 1998).

Conclusions and further research directions

SCC is increasingly becoming an organizational form even for technological development (Gulati, 1998; Das, 2000). Therefore, managers should develop their managerial skills needed to manage alliances. Companies might further reposition themselves as time passes, so it is crucial to evaluate the performance contribution of SCC to alliance firms as well as its own performance. The process of developing a performance measurement is not an easy task as the example of 3T shows.

Similar to the Turkish textile producers, many firms in developing countries are challenged to produce high value-added and innovative products instead of low quality and cheap products and export them to the world (Yoshino and Rangan, 1995). Our study firms target "producing high value-added and high quality products" and try to realize this through a collaborative network, namely the establishment of a joint venture. This is a successful collaborative venture with 3T as it enables the development of new technologies and products by bringing the resources of partners together. As well, 3T is a financially successful venture since it has become profitable after two years of its initiation. Such collaborations might be a tool for firms in developing countries in order to gain competitiveness in their respective industries at the global level. In fact, the success of Japanese firms is a good example of this (Chung and Kim, 2003).

On the other hand, it is not easy to build such collaborative networks and manage them well. This study attempts to identify the barriers to collaboration, where the lack of confidence to the partners and unwillingness to share information with the partners are of particular concern. This study also reveals that as the level of trust between the partners increases the performance of the collaboration and the benefits that the partners gain from the collaboration also increase. Besides trust, our study clearly shows that the existence of a performance measurement system is an extremely important element for an effective SCC.

Even though each alliance should attempt to develop its unique set of performance metrics depending on its specific characteristics, the case of 3T helps us to see how a performance measurement system can be applied where three metrics are used, namely (1) metrics related to internal performance, (2) perceptions of partners regarding the alliance performance, and (3) the degree of inter-firm relationship.

Overall, 3T is a successful venture. It satisfies its main goal by creating technologies that lead into the production of high value-added products at the industry. It is a profit making joint venture with expansion of new partners joining to the SCC. Its success is not limited to the creation of new technologies but also it covers the creation of positive perception of partners regarding the 3T's performance.

Even though 3T is successful, there are some potential improvement areas, particularly in its internal activities and its relationship with the partners. Internal performance can be further improved with small changes. For example, 3T should improve its performance either in estimating the duration of the projects more accurately or by completing the projects on time. It might improve its cost/benefit and payback analysis in its proposals since it tends to overestimate the reductions in labor costs, energy costs, and particularly in raw material costs.

The real benefit seems to come from the improvement of the degree of inter-firm relationship, in other words communication. Our study shows that as the level of information sharing and communication among the partners increase the performance and the benefits of the collaboration also increase. As the degree of inter-firm relationships affects the performance of SCC at a significant level, 3T should pay attention to develop its communication capability. This can be done in two ways. First, we observed that the involvements of the suppliers in the projects are not evenly distributed. The suppliers that take part in the projects less frequently may be included in the future projects by extending the scope of projects. More participation will help 3T build better integrated and higher value-added systems for its customers. Second, from the performance measurement analysis, we observe that the efforts of each partner to improve the collaboration are not at the same level. More frequent attendance of the partners to the meetings will also improve the communication and information sharing level among the partners.

It is not easy for SCCs such as 3T to measure its performance and evaluate its partners' performance in a comparative manner since the examples of such collaboration models are rare, preventing a benchmarking study. The continuous evaluations can bring revisions in performance systems, making it dynamic and flexible. For example, new performance metrics, especially in the area of resource sharing, may be developed when 3T extends its collaboration model to the area of resource sharing in the future.

One weakness of the study is its reliance on one joint venture for testing a performance measurement system. It would be useful to have more studies about technological collaborations in a supply chain. Moreover, having different technological collaboration forms might be another extension of this work that will help to compare the advantages of various governance structures besides joint ventures.

However, as the case of Turkish textile industry shows, SCC among small and large firms can be a successful tool in creating and diffusing innovation. Further, SCC can have long-term goals like continuous product innovation as experienced in 3T and this might increase overall performance of an industry, again as experienced in 3T (Chung and Kim, 2003). That is why SCC might be a particularly important tool for developing country firms that lack both financial and human resources. This in turn might increase the competitiveness of firms and the industry as a whole.

Managerial implications

- Technology collaboration is an advantageous strategy for competition, so it is of high concern to understand mechanisms behind the success of collaborations. Performance measurement system is one of these mechanisms that help not only to develop but also maintain collaborations.
- Considering the inherent complexity of the supply chain, selecting appropriate performance measures for supply chain collaboration is critical. Measures need to provide an accurate picture of supply chain performance as a whole but also highlight opportunities for improvement at the overall supply chain levels. Textile Dying Technologies (3T) is a collaborative model in the Turkish textile dyeing and finishing industry that struggles to establish a performance measurement system for effective supply chain collaboration. This is a joint venture of six technology

supplying and ten dyeing and finishing partner companies, each having a one-sixteenth share of 3T. It aims at developing and diffusing new technologies.

- Based on the analysis of 3T's practices, three critical performance measures are summarized below:

   (1) Managers should choose their internal performance metrics along the lines of innovation, sales, and productivity. Some examples are as follows:

   Innovation

   Percentage of the number of new products developed

   Sales

   Percentage of partner project proposals that are accepted by the alliance

   Percentage of alliance projects proposals that are approved by the partners

   Percentage of projects that are done for partners

   Productivity

   Average project time

   Percentage of projects that are completed on time

   Average time to respond to partners' project proposal

   Response time of alliance compared to other suppliers in the cases of breakdowns

   Frequency of project offers of alliance to the partners

   Average cost reduction percentage of alliance projects that is suggested by the alliance (labor/raw material/energy costs)

   Percentage of projects on which the price proposal has been revised

   Average suggested payback time of the projects.

   (2) Managers should find out perceptions of partners about the performance of alliance by measuring them with metrics such as:

   Comparison of the alliance performance to other suppliers

   Contribution of the alliance about the defect rates of the products that partners supply to the market

   Contribution of the alliance about the unplanned breakdowns of the machines of the partners

   Contribution of the alliance about the production capacity of the dyers

   Contribution of the alliance about the product quality of partners.

   (3) Managers need to identify solid measures to explore the degree of inter-firm relationship, including:

   Frequency of regular meetings

   Average participation percentage to the meetings.

References

Beamon, B. M. (1999). "Measuring Supply Chain Performance." *International Journal of Operations & Production Management* 19(3/4): 275–83.

Benfratello, L. and Sembenelli, A. (2002). "Research Joint Ventures and Firm Level Performance." *Research Policy* 31(4): 493–507.

Bititci, U. S., Mendibil, K., Martinez, V., and Albores, P. (2005). "Measuring and Managing Performance in Extended Enterprises." *International Journal of Operations & Production Management* 25(4): 333–53.

Bititci, U. S., Turner, T., and Begemann, C. (2000). "Dynamics of Performance Systems." *International Journal of Operations & Production Management* 20(6): 692–704.

Cahill, C., Charles, M., Fraser-Kraus, H., Boddy, D., and Macbeth, D. (1998). "Success and Failure in Implementing Supply Chain Partnering: An Empirical Study." *European Journal of Purchasing and Supply Management* 4: 143–51.

Caloghirou, Y., Ioannides, S., and Vonortas, N. S. (2003). "Research Joint Ventures." *Journal of Economic Surveys* 17(4): 541–70.

Carlsson, B., Jacobsson, S., Holmén, M., and Rickne, A. (2002). "Innovation Systems: Analytical and Methodological Issues." *Research Policy* 31(2): 233–45.

Chang, Y. -C. (2003). "Benefits of Co-operation on Innovative Performance: Evidence from Integrated Circuits and Biotechnology Firms in the UK and Taiwan." *R&D Management* 33(4): 425–37.

Chen, I. J. and Paulraj, A. (2004). "Towards a Theory of Supply Chain Management: the Constructs and Measurements." *Journal of Operations Management* 22: 119–50.

Childe, S. J. (1998). "The Extended Enterprise: A Concept for Co-Operation." *Production Planning and Control* 9(4): 320–7.

Chung, S. A. and Kim, G. M. (2003). "Performance Effects of Partnership between Manufacturers and Suppliers for New Product Development: The Supplier's Standpoint." *Research Policy* 32(4): 587–603.

Cooke, P. and Morgan, K. (1993). "The Network Paradigm: New Departures in Corporate and Regional Development." *Environment and Planning D: Society and Space* 11: 543–64.

Das, T. K. (2000). "A Resource-Based Theory of Strategic Alliances." *Journal of Management* 26(1): 31–62.

Dollinger, M. J. and Golden, P. A. (1992). "Interorganizational and Collective Strategies in Small Firms: Environmental Effects and Performance." *Journal of Management* 18: 695–715.

Dyer, J. H. (2000). "Creating and Managing a High-Performance Knowledge-Sharing Network: The Toyota Case." *Strategic Management Journal* 21: 345–67.

Dyer, J. H. and Singh, H. (1998). "The Relational View: Cooperative Strategy and Sources of Interorganizational Competitive Advantage." *Academy of Management Review* 23(4): 660–79.

Foreign Economic Relations Board (2002). *Turkish Textile and Apparel Sector.* Ankara: Foreign Economic Relations Board.

Freeman, C. and Soete, L. (1997). *The Economics of Industrial Innovation.* London: Pinter.

Gomes-Casseres, B. (1994). "Group versus Group: How Alliance Networks Compete." *Harvard Business Review* 72(4): 62–74.

Gulati, R. (1998). "Alliance and Network." *Strategic Management Journal* 19: 293–317.

Gunasekaran, A., Patel, C., and Tirtiroglu, E. (2001). "Performance Measures and Metrics in a Supply Chain Environment." *International Journal of Operations & Production* 21(1/2): 71–87.

Hagedoorn, J. and Cloodt, M. (2003). "Measuring Innovative Performance: Is There An Advantage in Using Multiple Indicators?." *Research Policy* 32(8): 1365–79.

Hagedoorn, J. and Schakenraad, J. (1994). "The Effect of Strategic Alliances on Company Performance." *Strategic Management Journal* 15: 291–309.

Harris, L., Coles, A. -M., and Dickson, K. (2000). "Building Innovation Networks: Issues of Strategy and Expertise." *Technology Analysis & Strategic Management* 12(2): 229–42.

Heide, J. B. and Stump, R. L. (1995). "Performance Implications of Buyer-Supplier Relationships in Industrial Markets: A Transaction Cost Explanation." *Journal of Business Research* 32(1): 57–66.

Johnson, D. W. and Johnson, R. T. (1989). *Cooperation and Competition: Theory and Research*. Edina, Minnesota: Interaction.

Kale, P., Dyer, J. H., and Singh, H. (2002). "Alliance Capability, Stock Market Response, and Long-Term Alliance Success: the Role of the Alliance Function." *Strategic Management Journal* 23(8): 747–67.

Kaplan, R. S. and Norton, D. P. (1992). "The Balanced Scorecard Measures That Drive Performance." *Harvard Business Review* (January/February): 71–9.

Kennerly, M. and Neely, A. (2002). "A Framework of the Factors Affecting the Evolution of Performance Measurement Systems." *International Journal of Operations & Production Management*, 22(11), p. 1222–45.

Kotabe, M., Xavier, M., and Domoto, H. (2003). "Gaining from Vertical Partnerships: Knowledge Transfer, Relationship Duration, and Supplier Performance Improvement in the us and Japanese Automotive Industries." *Strategic Management Journal* 24: 293–316.

Lambert, D. and Cooper, M. (2000). "Issues in Supply Chain Management." *Industrial Marketing Management* 29(1): 65–83.

Landeros, R., Reck, R., and Plank, R. E. (1995). "Maintaining Buyer-Supplier Partnerships." *International Journal of Purchasing and Materials Management* 31(3): 3–12.

Lapide, L. (1998). *What About Measuring Supply Chain Performance*. Advanced Manufacturing Research.

Malecki, E. and Tootle, D. (1996). "The Role of Networks in Small Firm Competitiveness." *International Journal of Technology Management* 11: 43–57.

Medcof, J. F. (1997). "Why too Many Alliances End in Divorce." *Long Range Planning* 30(5): 718–32.

Neely, A., Gregory, M., and Platts, K. (1995). "Performance Measurement System Design." *International Journal of Operations & Production Management* 15(4): 80–116.

Olk, P. (1997) "The Effect of Partner Differences on the Performance of R&D Consortia." In P. W. Beamish and J. P. Killing (Eds), *Cooperative Strategies: Vol. 1 North American Perspectives*. San Francisco, CA: Lexington Press: 133–59.

Pagh, J. D. and Cooper, M. C. (1998). "Supply Chain Postponment and Speculation Strategies: How to Choose the Right Strategy." *Journal of Business Logistics* 19(2): 13–33.

Ramirez, R. (1999). "Value Co-production: Intellectual Origins and Implications for Practice and Research." *Strategic Management Journal* 20: 49–65.

Reuer, J. J. and Miller, K. D. (1997). "Agency Costs and the Performance Implications of International Joint Venture Internalization." *Strategic Management Journal* 18: 425–98.

Sako, M. and Helper, S. (1998). "Determinant of Trust in Supplier Relations: Evidence from the Automotive Industry in Japan and the United States." *Journal of Economic Behavior & Organization* 34: 387–417.

Salcedo, S., and Grackin, A. (2000). "The e-Value Chain." *Supply Chain Management Review* (Winter): 63–70.

Saxenian, A. (1994). *Regional Advantage: Culture and Competition in Silicon Valley and Route 128*. Cambridge, MA: Harvard University Press.

Staber, U. (1996). "Networks and Regional Development: Perspectives and Unresolved Issues." In Staber, U., Schaefer, N. and Sharma, B. (Eds), Business Networks: Prospects for Regional Development. Berlin: Walter de Gruyter.

Tan, K. C. (2002). "Supply Chain Management: Practices, Concerns, and Performance Issues." *Journal of Supply Chain Management* 38(1): 42–54.

van Hoek, R. I. (1998). "Measuring the Unmeasurable: Measuring and Improving Performance in the Supply Chain." *Supply Chain Management* 3(4): 187–92.

von Hippel, E. (1988). *The Sources of Innovation*. Oxford: Oxford University Press.

Womack, J. P., Jones, D. T., and Ross, D. (1990). *The Machine that Changed the World*. New York: Rawson Associates.

Yoshino, M. Y. and Rangan, U. S. (1995). *Strategic Alliances: An Entrepreneurial Approach to Globalization*. Boston, MA: Harvard Business School Press.

Part III
Innovation

9

Understanding Discontinuous Technology and Radical Innovation

Gaston Trauffler and Hugo Tschirky

Introduction

There is a great variety of terms used to describe the phenomenon of technological discontinuity and radical innovation (Green et al., 1995: 203; Garcia and Roger, 2002: 110). Both terms are often used synonymously without any differentiation (Lehmann, 1994: 10). Besides the term radical innovation (North and Tucker, 1987: 11; Damanpour, 1988: 546; Leifer, 2000: 4; O'Connor and Veryzer, 2001: 233), and breakthrough innovation (Nayak and Ketteringham, 1986: 181; Mascitelli, 2000), authors speak of breakpoint (Strebel, 1995: 11) or revolutionary innovation (Abernathy and Clark, 1991: 61). Since a few years Christensen (1997; 2003) introduced the term disruptive technology in the context related to radical innovation changes. The present article will focus on the most popular terms used in literature – which are discontinuous technology, radical innovation, and disruptive technology – in order to differentiate them and to show how they are related. Doing so, it will also shed light on how the research community commonly understands the phenomenon described by these terms.

Based on this common understanding scholars derived a set of generally accepted implications that describe the challenges related to the management of discontinuous technology and radical innovation.

General understanding

Terminology and phenomenon

A discontinuous technology is a technology that is the result of a scientific breakthrough achieved through a long-term research effort. It builds on newly acquired scientific knowledge that usually does not follow the expected evolution of existing technologies. It is the deviation from what was expected by forecasting experience that makes the technology discontinuous (Lehmann, 1994: 8ff). Thus, it represents a breakpoint within a given technological paradigm[1] that initiates a change to a new technology trajectory (Dosi, 1982: 152). Tushman and Anderson (1997: 5) point out that technological discontinuities can be competence enhancing as well as competence destroying.[2] As a consequence, not all discontinuities affect companies in the same way. For some companies the same technological discontinuity might be helpful to sustain competitive advantage as the discontinuity enhances its existing competencies while for the other company it might mean the debasement and destruction of its competencies (McKelvey, 1996: 108ff). As a consequence, this company has to substitute the greater part of its technological knowledge in order to follow the discontinuity. Thus, there are discontinuities that affect only a few companies while other discontinuities affect a whole industry.

From a market perspective discontinuous technologies often provide the basis for a new product with an order of magnitude improvement (actual or potential) in price/performance[3] ratio, caused by a substantially changed technology[4] base (Anderson and Tushman, 1990: 604; Ehrnberg and Jacobsson, 1993: 28).

The successful commercialization of a discontinuous technology can lead to either a radical innovation or to an incremental innovation (Brodbeck et al., 2003: 137). Concretely speaking a radical innovation involves the (1) application of significant new technologies often emerging from a discontinuity, or (2) significant new combinations of existing technologies to new market opportunities (Tushman and Nadler, 1986: 74f). Radical innovation "departs dramatically from the norm" (Anderson and Tushman, 1990: 604) and "transforms the relationship between customers and suppliers, restructures marketplace economics, displaces current products, and often creates entirely new product categories" (Leifer et al., 2000: 2). It can lead to either a new product class life-cycle, or a discontinuity in an existing life-cycle, initiating the substitution of an old product or process for the new one (Lambe and Spekman, 1997: 102). Radical innovations bring forth whole product lines that are not only new for the company that creates it but that are also new for the market place. Often such radical innovation addresses an urgent customer need that had yet not been articulated, or it noticeably improves the price/performance ratio of existing products (O'Connor, 1998: 151).

Beside these definitions of radical innovations that are primarily focused on a market perspective, a few authors only give definitions or rather descriptions of radical innovations from a company perspective. Green et al. (1995: 203ff) for example consider innovations radical if the four following dimensions apply: (1) technological uncertainty, (2) technical inexperience, (3) business inexperience, and (4) high technology cost. Shenhar et al. (1995: 179) classify technologies according to their uncertainty at the time the project is initiated at the company. With regard to the fact that most technological projects employ a mixture of technologies, some are emergent and highly uncertain and some more mature and less uncertain he suggests defining four types of innovations according to the overall uncertainty of the project: low-technological uncertainty over medium- to high-up to super-high technology uncertainty.

Similarly to radical innovations, disruptive technologies address the commercialization of a discontinuous technology. As a matter of fact technological discontinuities are considered a subset of technological disruptions (Ehrnberg and Jacobsson, 1993: 44). A discontinuity is considered a disruption if it has a restructuring effect on an established industry (Christensen, 1997: 13ff; Porter, 1998: 253; Rafii and Kampas, 2002: 116; Vojak and Chambers, 2004: 123). Ehrenberg and Jacobsson (1993: 44) hypothesize three causes that might play a role when a discontinuity turns into a disruption. As a first cause they suggest the destructive character of the technology. The more the discontinuity destructs or renders obsolete existing technological knowledge, the more it might have a disruptive effect on the industry. Second, the more new technologies from a distinct generic area are added to the existing technology base of the industry, the higher the probability for a disruption. As a third cause, these authors name the rate of diffusion of the technology. It is stated that the faster the diffusion of the discontinuity the more disruptive its effect will be. This is due to the fact that the time companies have at their disposal to detect and to react to a discontinuity is determined by the speed of diffusion of that technology between users.

Although disruptive technologies are seen to be a threat to an established industry they usually create growth in those industries they penetrate. They might even create entirely new industries by introducing new products and services (Kostoff et al., 2004: 141). Typically, disruptive technologies emerge in industries different from the ones they finally disrupt. In their initial industry they first appear in niche markets from which they grow to penetrate other industries with more attractive markets in terms of margin and volume (Christensen, 1997; Gilbert, 2003: 28).

The above-mentioned definitions of disruptive technologies and discontinuous technology focus mainly on changes and the consequences of these phenomena from a product, market, and customer view within the industry as a whole[5] and within a company in particular. In sum, a discontinuous technology or a radical innovation initiates major changes within a given structure of an industry while a disruptive technology triggers changes that restructure a whole industry. Descriptions and definitions of radical innovations are also considered from a market view, based on their degree of newness compared to existing products or through their price/performance competitiveness. This most common understanding of discontinuous and disruptive technologies and radical innovation is emphasized in Table 9.1 adapted from Kassicieh et al. (2002: 385). It shows from which most common perspectives different authors define the three terms. These perspectives are related to a specific characteristic of change – (1) change in technology, (2)

Table 9.1   Four perspectives of defining discontinuous/disruptive technologies and

	Corroad 1982	Abernathy & Clark 1985	Foster 1986	Meyers & Tucker 1989	Moore 1991	Mckee 1992	Bower & Christensen 1995
Perspective 1: change in technology							
Change in technology base							
Technology learning curve						•	
Perspective 2: change in technology and products							
Technology / product paradigm						•	
New industry architectural technology / product paradigm		•					
New industry revolution technology /product paradigm		•					
New product families							
Perspective 3: change in markets							
Newness of technology / product / market	•						
New firm technology/ product / market paradigm							•
Perspective 4: change in customer benefits							
Change in customer benefits					•		
Order of magnitude improvement in cost or performance							
Increase in user benefits				•			
Competitive advantage			•				

Source: Adapted from Kassicieh et al., 2002.

radical innovations

Ehrenberg 1995a	Lynn, Morone & Paulson 1996	Walsh 1996	Lambe & Spekman 1997	Veryzer 1998	Rice et al. 1998	Leifer et al. 2000	McDermott et al. 2002	Clark 2003
			●					
	●					●		
●				●			●	●
						●		
		●			●			
				●				

change in technology and corresponding products, (3) change in markets, and (4) change in benefit of the customer – through which authors define discontinuous/disruptive technology or radical innovation. It is apparent from this Table 9.1 that most authors focus on perspectives 2, 3, and 4.

Implications: challenges for management

With this basic understanding being set, this section presents the implications of the above described phenomenon for the management of a company. The implications shown in this chapter represent the common understanding within the whole community of researchers – industry focused and company focused – that are investigating in this field.

There is a broad agreement among researchers that management usually handles the emergence of discontinuous technologies and radical innovation with a low sense of urgency (Lambe and Spekman, 1997; Rafii and Kampas, 2002: 102). This is due to the long-term perspective of the research and development projects of these technologies. Generally such projects last up to ten years and more (Rice et al., 1998: 58) which makes it difficult for managers to see the applicable outcome of these efforts right from the beginning. Thus, such projects are often run with low priority having a minor sense of urgency. However, these projects require a great deal of resources and management attention (McDermott and O'Connor, 2002: 425) before they can eventually be transferred into marketable products. These contradictory characteristics – low priority and minor sense of urgency compared to the high degree of resource and management requirement – results in conflicts. Furthermore, the progression of radical innovation projects starting from basic research via development to the first stages of commercialization is an endeavor accompanied by a high level of uncertainty (Abernathy and Utterback, 1978: 45; Rice et al., 1998: 58; Veryzer, 1998a: 318). Such uncertainty is of multiple dimensions (Milliken, 1990; Leifer, 2000: 18ff). For instance, it is due to a lack of technological and market knowledge as no previous technological or reliable market insight exists in the company (Christensen, 1997: 209; Jolly, 1997: 7ff). Market data is seldom available and customer requirements are often vague[6] (Wieandt, 1995: 450; Song and Montoya-Weiss, 1998: 132; Veryzer, 1998: 149). Veryzer (1998: 147) brings it to the point: "for discontinuous new products customer input may not necessarily be relied upon as heavily to guide the product development process as it is in developing incremental new products." All of this lack of information leads to uncertainty making strategic planning of radical innovation projects very difficult.

It is the people within the existing organization, meaning managers, that are most concerned by the uncertainty inherent to a discontinuous technology or radical innovation project. Often managers have difficulties seeing the business opportunity targeted with such projects as they do not fit in their existing mindset formed through the present portfolio of projects. Many companies view business opportunities too narrowly through the lens of their existing assets and capabilities (Kim and Mauborgne, 1997: 106). This cognitive filter is referred to as dominant logic. It is the set of biases, beliefs, and assumptions – about the markets to enter, technologies to use, competitors to watch, people to hire, and businesses to run – that is rooted within an organization (Afuah, 1998: 97ff). Dominant logic is defined as "a mind set or a world view or conceptualization of the business and the administrative tools to accomplish goals and make decisions in that business. It is stored as a shared cognitive map (or set of schemas) among the dominant coalition" (Prahalad and Bettis, 1986: 491). The dominant logic of an existing company hinders organizations from recognizing the opportunity behind a totally new technology. Thus, it may provide at least a partial explanation for the uncertainty organizations encounter when confronted with discontinuous technology or radical innovation.

In addition to the existing organizational hurdles described above there are resource uncertainties that complicate the decision for and the implementation of radical innovation projects. Resource uncertainties are the result of the probability of a major loss of funding because of an overall decrease in corporate performance or a change in senior management sponsorship (Leifer, 2000: 23). In sum many of these uncertainties coupled with a high level of resource assignment make discontinuous technology and radical innovation projects very risky. This is why many organizations are reluctant to engage in such projects (McDermott and O'Connor, 2002: 425) and tend rather to further develop their competencies within a relatively narrow scope and range (McKelvey, 1996: 109) focusing on short-term revenues. Thus, once a discontinuous technology is ready for the market, it is often commercialized by outsider companies instead of established industry leaders (Utterback, 1994: 160; Strebel, 1995: 11; Christensen, 1997: 85).

In general, there is broad agreement between scholars that discontinuous technologies and radical innovations have a very specific character that is distinct from continuously evolving technologies and incremental innovation (Lynn et al., 1996: 11; O'Connor, 1998: 162; Rice at al., 1998: 57; Song and Montoya-Weiss, 1998: 132; Kessler and Chakrabarti, 1999: 231). This distinction of technologies and innovations asks for different styles of management including differentiated types of strategic actions and organizational capabilities (Kessler and Chakrabarti, 1999: 235). Thus, conventional management techniques are not suitable until the technological innovation has reached a certain maturity level so that it can fit the pattern of incremental innovation (Abernathy and Clark, 1985: 20; Rice et al., 1998: 58; Veryzer, 1998a: 319; Kessler and Chakrabarti, 1999: 231; Leifer et al., 2000: 11). Authors (Tushman and Anderson, 1986; Tushman and O'Reilly, 1996a; McDermott and O'Connor, 2002) agree that sustainable growth requires specific management skills for both types of innovation – incremental and radical – at the same time. Thus, Tushman and Anderson claim (1986: 734) that companies have to overcome the dilemma to master "evolutionary and revolutionary change" simultaneously. Tushman and O'Reilly (1998: 40) emphasize that on the one hand, companies have to plan and align their activities along a relatively stable and evolutionary change. On the other hand, they have to eliminate these achievements once the competitive environment changes radically knowing that new technologies will substitute the foundations underlying their present products.

These understandings that reflect the very different nature of radical versus incremental innovation represent the initial positions for all researchers active in this field. They commonly acknowledge that the inherent uncertainty and risk of radical innovation needs a distinguished management from the one used for incremental innovation. However important this distinction is, scholars agree that the dilemma lies in the necessity for the simultaneous management of both types of innovations. This simultaneous consideration is called sustained innovation management, it manages the combination of changes initiated by radical innovation and followed by continuous incremental improvement innovation. The dilemma is initiated by the requirement to fulfill conflictive criteria in one and the same innovation management concept. Thus, implications for management are not intuitively clear. It needs to adopt a "sustained innovation" perspective. This perspective requires following three criteria:

The capability to simultaneously manage long-term and short-term issues in differentiated ways.

The capability to simultaneously manage issues with different levels of risk in differentiated ways.

The capability to align management issues of long-term and shot term character, as well as management issued with different risk levels to one single strategic planning purpose.

Taking the common understanding of this section as a basis the next sections will show different solution approaches suggested by the two clusters of research – industry focused and company focused – mentioned earlier.

Solution approaches

Building upon this overview the present section will analyze which solution approaches scholars suggest that contribute to successful management of discontinuous/disruptive technologies and radical innovation. Thus, the two clusters of research – industry focused and the company focused – will be analyzed and presented in more detail. Some of the authors that will be presented have already been mentioned above in the context of description and definition of the phenomenon and its implications for management. These authors' research will now be presented from a solution-providing perspective.

Industry level focused research

Research focused on the industry level builds on the common understanding of discontinuous / disruptive technologies and radical innovation. Starting from this understanding it aims to describe the consequences of a technological discontinuity on the macro economic level within a given industry structure. Thus, scholars collect empirical data over long periods of time within one or more specific industries. Analyzing this data, they look for patterns that emerge whenever a discontinuity happens. The benefit of this research is twofold: on the one hand, it allows the dynamics of the phenomenon of discontinuous technologies to be studied and on the other hand, modeling and generalizing observed patterns helps to develop planning approaches for the management of future discontinuities.

First a group of researcher's work will be presented that focuses on developing life-cycle models out of their observations. In a second part of this section a group of researchers' work will be presented that, based on their observed pattern directly discuss approaches and models describing how to handle the observed pattern from a generalized management perspective, finally a market commercialization perspective will be presented.

Life-cycle perspective

A first well-known life-cycle model is the technology S-Curve. First presented by Foster (1986) it visualizes how technology performance evolves with cumulating R&D expenses (or time) along an s-shaped curve (see Figure 9.1) (Foster, 1986: 31). The performance growth curve in a first phase is low as the technological progress of an emerging technology is typically slow. However, with accumulating expenditures in R&D this performance can be continuously increased. In this second phase the R&D yield per invested unit is high due to a good basic knowledge. In its third phase the model of the technology performance curve shows a decreasing increment, approaching a certain limit. The curve flattens increasingly down as the model is based on the assumption that every technology has a certain performance limit that cannot be exceeded. This limit often represents the end of this technology as further R&D investments will no longer increase the technologies' performance significantly and eventually R&D investment will be stopped in favor of an a new technology with the potential to exceed the performance of the old technology. At this moment the S-curve of this old technology stops, it is discontinued. Technology that is now invested in, is called "discontinuous" compared to "the old technology"; it initiated the discontinuity of its predecessor technology.

Analyzing Foster's work from the perspective of its contribution for the purpose of this research there are two criteria fulfilled: first the requirement for a strategic perspective and second the requirement for supportive tools and techniques. As a matter of fact Foster promotes the S-curve as a strategic planning tool. He argues that it can forecast when and how fast a technology will attain its performance limits. Following the shape of the curve would allow the prediction to of which extent present products can be improved and what their cost would be. However some scholars disagree with the usefulness of the S-curve as a strategic planning instrument (Osterloh and von Wartburg, 1998: 141). There are two main augments why the contribution of the S-curve for strategic planning is limited: first the actual progression of the technology performance curve is often not in the course of a predicted S-curve. It is especially the limit of the technology performance that is hard to forecast with accuracy. Many cases showed that the technology that is predicted to be discontinued would at the moment of distress boost its performance to a higher limit than had been expected. Such unexpected performance boosts at the end of a technology S-curve are described by the term "sailing-ship effect" (Foster and Kaplan, 2002: 140ff). A second criticism of the S-curve is that it is in the first place of a descriptive nature. It does not give any strategic recommendation of how to proceed in a given situation. It is particularly criticized that the concept does not show any relationship between technological evolution and product/market opportunities. Taking these two points of criticism into account, Foster's contribution for the purpose of this research is limited.

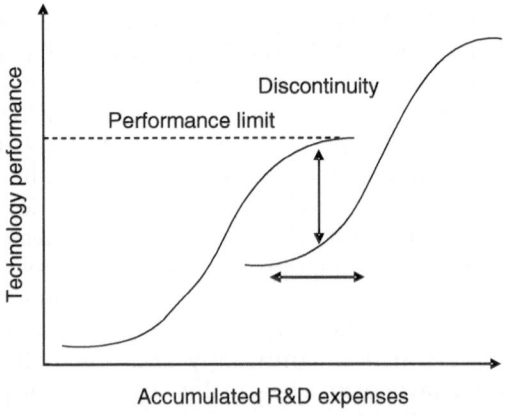

Figure 9.1   Technology S-curves
Source: Foster, 1986: 111.

A second model is the technology lifecycle model.[7] It is based on the observation that over time, technologies run through different stages of market diffusion similar to the product life cycle model. Typically four stages can be differentiated. The first stage describes the emergence of the technology, it is characterized "by basic discovery, scientific turbulence and highly uncertain potential" (Roussel, 1984: 29). During this stage the technology is called pace maker technology. In the second stage of market penetration, the technology is referred to as key technology. This stage is characterized by increasing diffusion until the third stage of maturity begins, where the technological advance slows down. In this stage the technology is referred to as basic technology. In its last stage, the technology slowly disappears from the market. This stage is called degeneration or the aging stage of the now named threatened technology. According to experience the technologies' competitive capacity declines with progression through its lifecycle. While the progression through the lifecycle is driven by incremental innovation, the emergence of a new technological lifecycle will be generated by a radical innovation (see Figure 9.2).

The use of the technology life cycle in strategic planning is similar to the S-curve model. Theoretically the visualization of two successively following technology lifecycles can indicate the emergence of a technological discontinuity. However, it is generally hard to predict how market diffusion of a technology will be over time. Thus, the technology life cycle model is purely descriptive and is difficult to use for strategic planning purposes.

A third model is the industry lifecycle model. It originates from the observation that structures and competition within many industries are influenced to a great extent by the technological paradigms applied in it (Abernathy and Utterback, 1978: 43; Abernathy and Clark, 1985: 13). Thus, a discontinuous change of a technology paradigm, even initiated by a single

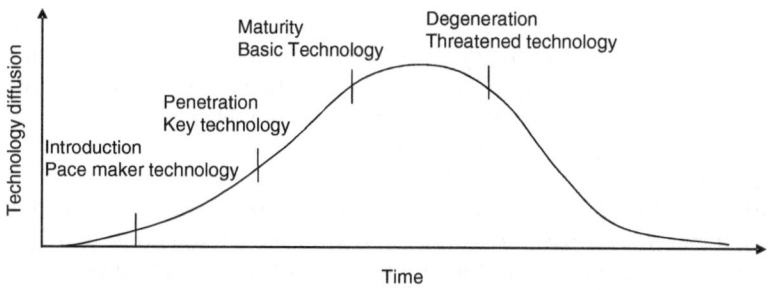

Figure 9.2 The technology life cycle

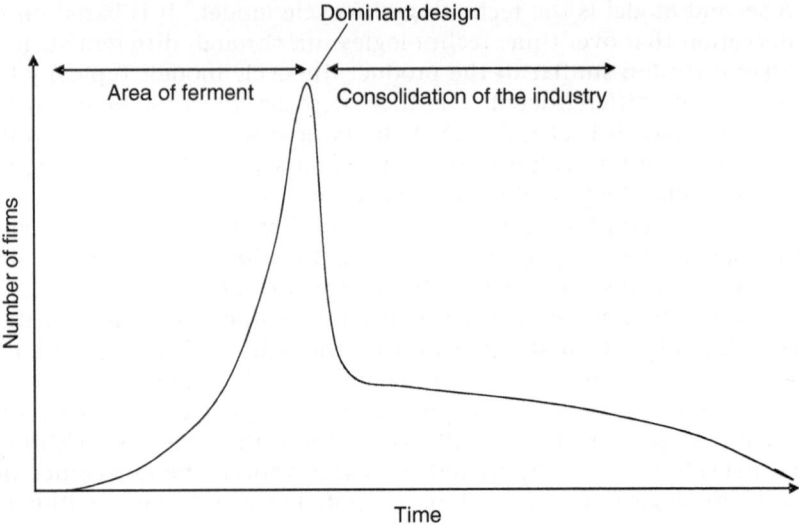

Figure 9.3   Typical course of the industry life cycle

company, has considerable affect on the whole industry (Tushman and Anderson, 1986: 460).

The observed pattern is that a radical innovation enlivens the innovation rate in an industry during the so-called area of ferment which is characterized by two distinct selection processes: "competition between technical regimes and competition within the new technical regime" (Anderson and Tushman, 1990: 606). In this competition, the number of innovations and the number of firms competing in the industry rises until a so called dominant design emerges that establishes a new technical regime that is most accepted in the market (see Figure 9.3). The dominant design focuses technology change on a specified and widely recognized incremental improvement trajectory (Stoelhorst, 2002: 265). With the dominant design established, a new technology cycle begins (see Figure 9.4). The number of innovations competing decreases and the industry consolidates, which means that number of companies in the industry decreases. During the creation of the dominant design, the industry is usually totally restructured. New players enter in the industry; established ones vanish, with them the customers' constellation alters as well as the set of suppliers to the industry changes.

Although this model is just as descriptive as the previous two, it has the power to serve as a basis for strategic planning. There is one main difference in this model compared to the two previously described ones, that is, it does not make any statement that is critically time dependent. While the previous models described one single pattern which only becomes valuable

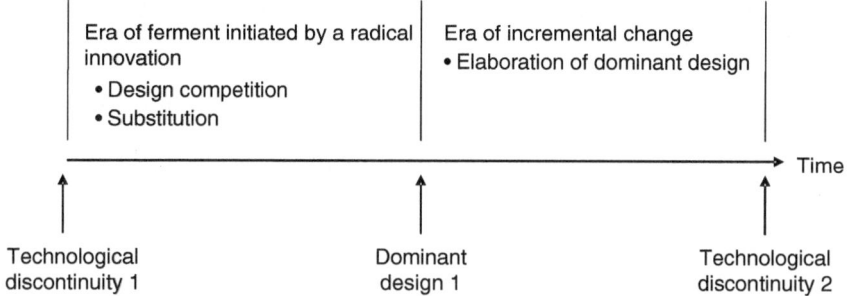

Figure 9.4 The industry technology cycle
Source: Anderson and Tushman, 1990: 606.

when considered in the course of time, the present model describes the alternation of two successive patterns over time. It is the statement that two patterns alternate whenever a technological discontinuity happens that is valuable for strategic planning. Now companies know what pattern they have to be prepared for although they still do not exactly know when these preparations need to be ready. As a consequence, management approaches are developed in order to describe how to prepare companies for discontinuous change that now make it clear what pattern to prepare for.

The first management approaches developed in the course of this research date from 1985 (Abernathy and Clark) and 1986 (Tushman and Anderson). Based on the observations made on industrial change as a consequence of technology change, these researchers claimed that there are differentiated organizational environments needed according to the type of technology change encountered. Later in 1990 (Anderson and Tushman) and 1994 (Utterback) when the above shown model was developed and technology change was related to an alternating pattern within the industry the requirement for approaching discontinuous technology change became more distinct Anderson and Tushman (1990: 629) claim that organizations "must develop diverse competencies both to shape and deal with technological evolution." First, firms have to build up capabilities to either initiate or respond rapidly to discontinuities. Second, inter-organizational dynamics are required meaning that organizations must be able "to combine technological capabilities with the ability to shape inter-organizational network and coalitions to influence the development of industry standards" (Anderson and Tushman, 1990: 629). Finally, companies need to master the ability to produce incremental innovations for the time after the dominant design has been established. These three approaches has been further developed and adapted to a more strategic perspective. In Tushman and O'Reilly (1996), Tushman and O'Reilly (1998) and recently in 2004 O'Reilly and

Tushman they were published under the approach named "Ambidextrous Organizations." The main message of these publication being: "The real test of leadership [...] is to be able to compete successfully by both increasing the alignment or fit among strategy, structure, culture, and processes, while simultaneously preparing for the inevitable revolutions required by discontinuous environmental change. This requires organizations and management skills to compete in a mature market (where cost, efficiency, and incremental innovation are key) and to develop new products and services (where radical innovation, speed, and flexibility are critical" (Tushman and O'Reilly, 1996a: 726).

Evaluating the above course of research with regards to the requirement for a solution approach targeted in the present research context, at least two requirements are fulfilled. There is a distinct strategic perspective recognizable in the solution approaches described as well as a sustained innovation perspective promoting simultaneous management of radical and incremental innovation. To some extent there are also statements that include the structural aspects of the organization, although these statements generally remain quite vague.

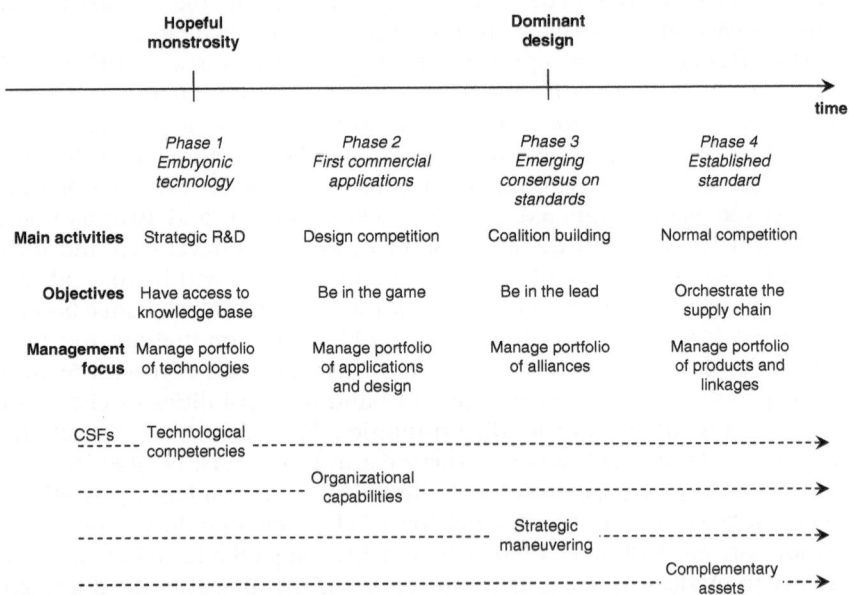

Figure 9.5   Management tasks along the evolution of a radical new technology
Source: Stoelhorst, 2002: 278.

A forth model is developed by Stoelhorst (2002: 278ff). He models the process by which a radically new technology evolves through different phases into a regime that is, subsequently only incrementally improved. As an initial position to an innovations' evolution process, he uses the term of "hopeful monstrosity" adopted from Mokyr (1990). In terms of Anderson and Tushman the appearance of a hopeful monstrosity slowly initiates the area of ferment. It is the moment in time when a first application of a technology is visible however it is still in the phase of a "crude and expensive principle of device" (Stoelhorst, 2002: 264). Along the process started by the hopeful monstrosity, he describes the competitive effects of discontinuities observed in an industry according to four phases. For each of these phases he further describes management's main activities, its objectives, and its focus. For each of the phases he additionally indicates critical success factors (CSFs) that need special emphasis (see Figure 9.5).

Analyzing Stoelhorst's contribution from the beneficial perspective for this research it can be said that the described model and especially the management activities assigned to each of the phases respond to two requirements: the one calling for a planning process and the other for a strategic perspective. However the tasks enumerated under each of the phases do not yet result in a process although their structure and assigned listings bring some clarity in how a process could be approached systematically.

Christensen (1997) models his macroeconomic finding on the dynamics of the hard disc industry using the S-curve (1997: 41). Analyzing how the hard disc industry changed over a period of 40 years, he then elaborates why companies that were well managed and positioned at the top of the industry failed when faced with a discontinuous/disruptive technology change. Based on this analysis he then elaborates the four main reasons for this failure (Christensen, 1997: xix ff). The first one is that established companies are sticking too much to their known and existing technologies. Christensen calls them "sustaining technologies." The main goal of this approach is to improve and sustain proven products for the mainstream markets in order to achieve, through learning and economies of scale the revenue that is expected from large, well established companies. Concentrated fully on such sustaining technologies these companies do not pay attention to the emergence of a discontinuous/disruptive technology and the need to change their sustaining technology in favor of a new one. The second reason, as consequence of the first approach is that for established companies that are looking for growth, the small markets that are typically opened by emerging technologies are not attractive. The market size is insufficient for a big company to invest in it. Third, is that emerging markets are hard to assess. Relying on conventional market research, good planning can assess growth rates of existing markets as customer needs are known and trajectories of technologies can be estimated.

However, these same approaches do not fit when assessing markets created by disruptive technologies, as there is no historical data available on such emerging markets. This makes its assessment extremely difficult. As a fourth reason, Christensen indicates the technological supply overshooting the present market expectation. This is due to the fact that established companies in their efforts to increasingly improve existing products in order to stay competitive might overshoot present market expectation in terms of price/performance causing customers to look for alternative products with a lower level of performance and price. Often this is the opportunity for disruptive technologies to enter the market very rapidly as they typically provide lower performance at reduced prices. Their inferior price/performance characteristics that formerly used to handicap disruptive technologies from entering the market may become their argument for competition.

Christensen summarizes all these reasons that cause well managed companies to stumble over discontinuous technology under the term of the "innovator's dilemma." Although the dilemma as such is well described, solutions to it are scarce. Some approaches to solving it are developed using the case of an electronic vehicle as a disruptive technology (Christensen 1997: 187ff). The most applicable approach, however presented entirely from an operational perspective, is an organizational one. Recommended for the commercialization of a disruptive technology in an established company is the creation of an independent organization spin off from the established mother company. Three main justifications for this recommendation can be interpreted from Christensen's explanations. First an independent organization spin off from the mother company is more resource independent. Second, it allows a small sized organization to exploit an equal sized market without having to respond to revenue expectations of its mother company and third, as an independent organization it is more likely to be able to create an appropriate attitude to failure.

Generalized management perspective

There are a number of authors such as Christensen and Raynor (2003), Clark (2003), etc. (see Table 9.2) in the industry-focused research that handles the challenge of discontinuous technologies from a generalized management perspective. Based on their finding they describe management approaches that in a generalized way apply to managing the subject of their research. The range of aspects covered by these researchers is broad and heterogeneous. In order to provide a structured overview nevertheless, the following section structures the different aspects of generalized management perspective from an industry focused research along the six generic strategy development process steps often used in general management which are (1) definition of strategic objectives, (2) environmental analysis, (3) Company

Table 9.2 Generalized management perspective of industry level focused research and its contribution to the generic strategy development process

Authors	Year	a. Strategic objectives	b. Environmental analysis	c. Company analysis	d. Strategic options	e. Strategies	f. Strategy implementation
Strebel	1992	■	■		■	■	■
Bower & Christensen	1995		■				■
Bower & Christensen	1995a		■	■			■
Suarez & Utterback	1995		■				
Floyd	1996			■	■	■	
Kusunoki	1997					■	
Song & Montoya-Weiss	1998	Global statements over whole process		■			■
Clark & Bower	2002						■
Clark	2003		■				
Christensen & Raynor	2003					■	■

analysis, (4) evaluating strategic options, (5) strategy formulations, and (6) Strategy implementation.[8] Such a structure leads to a table that opposes the six steps of the generic strategy development process to all but one of the authors presented in this section (see Table 9.2). The assignment of an author to a step in the strategy process shows the author's general contribution to that specific step of strategy development.

Strategic objectives

Strebel (1992: 227ff) is the only author who explicitly describes aspects of strategic objectives in the context of managing discontinuities. He points out that those organizations with the capability of initiating discontinuities are often oriented towards long-term objectives that exceed their present competencies and expertise. Strong strategic goals for such organizations serve as guidelines in order to achieve their objectives in competence extension. These goals, if well formulated, provide companies with set benchmarks along which continuous change of capabilities is oriented in order to create options that have the potential to trigger discontinuities. Besides its importance in strategic management, strong goals that are well communicated in an organization help to make people aware that all changes are focused in one direction. These goals stand for top management's persistency and resolution. This is an aspect not to be neglected, as in the case for most people; changes are generally encountered with skepticism and evoke uncertainty.

Environmental analysis

The environmental analysis is quite well described in the context of discontinuous technologies. Generally speaking, all five authors handling this aspect of the strategy development process recommend companies to open their mind wider to look for signs outside their usual fields of interests. Strebel (1992: 25ff) describes a procedure of three steps to identify the dynamic in the environment. First to assess are the forces for change. By these forces general trends, cyclical pattern and emerging turning points in the political, technological, economic, and social environment are meant. Second is to assess the forces of resistance. Such forces can be found in structures and systems within the organization, within the industry, with the stakeholders or generally in society. They usually reflect the impact of change on values and to what extent behavior and skills have to adapt to the forces of change. Examining to what extent the various forces of resistance correlate can assess the overall strength of resistance. Third is to build scenarios based on the analysis from steps one and two. Such scenarios can map what the consequences will be when forces of change and forces of resistance evolve and eventually collide. They help to identify which part of an industry is most likely to change.

Floyd (1996: 8) also suggests a scenario approach. He applies scenarios to a more narrowed analysis focused on the evolution of the market. These scenarios should be looking into a far distant future that is 20 years and more ahead. It should build on extrapolations of today's market trends and give an image of how a distant future might look. The accuracy of such scenarios is usually weak in hindsight; however their main purpose is not to predict the future but to sketch a desired one that helps the company to imagine and discuss probable futures. Even if accuracy in scenarios is not a primary goal, including customer views in the scenarios might give them a false image as customers tend to think in the frame of a present view (Bower and Christensen, 1995; 1995a: 93). This present view is often too limited to correctly assess technological discontinuities that might emerge in the distant future. The customers of such a future will probably have totally different needs from the ones today. Thus, the identification of future customers is not an easy task, as the products that they want to buy are not yet imagined, not to mention that the need they want to have satisfied through them is often not even articulated. Clark (2002: 29) based on finding from Christensen et al. (2002) suggests that in order to find new customers the new markets have to be identified first. He proposes to check three criteria in order to find out whether a radical innovation has potential for a market. First, the innovation "must be undervalued by current customers. Second, it must compete against non-consumption, that is, it must allow people to do things they could not do in the past due to a lack of money or skill. And, third, it must help people accomplish things that they are already trying to do but can not with the available products or services" (Gilbert, 2002: 31). Once those new customers are found it is important to stick with them to learn how to keep them by providing them with solutions driving improvement in the way of their changing needs. This is often only possible if new business models are developed, as most discontinuous technologies open possibilities that cannot fully be studied with existing business approaches.

Company analysis

In the step "company analysis" the discontinuous technology is strategically analyzed in relation to the company. According to Floyd (1996: 5ff) it is important here to assess first the company's internal vulnerability. This assessment helps to detect which of the internal technologies are running out of "steam" and are most likely to be substituted with an emerging discontinuity. Bower and Christensen (1995) suggest similar procedures: A check of exactly what threat and what strategic meaning a specific discontinuous technology represents for the company, is suggested, extrapolations of technology performances and tracking of technology S-curves are suggested as supporting management tools for these considerations.

Objectively seen, the action proposed in the previous paragraph does not differentiate much from a well-known and generally accepted strategic approach, that of the SWOT analysis developed by the General Management group at the Harvard Business School and published in the basic textbook *Business Policy: Text and Cases* by Learned et al. (1965). Thus, the approaches presented so far, which contribute to "company analysis," have only limited character that is specific to the management of discontinuous technologies.

More specific are the propositions from Clark and Bower (2002: 4ff), which note that, in order to direct the necessary management attention to an upcoming radical innovation it is best framed first as a threat. Their findings indicate that the way a challenge is perceived – in this case a discontinuous technological innovation – influences an organization's behavior when approaching it. When a radical innovation is only seen as an opportunity, resources are allocated too scarcely. "Framing the innovation as a threat will generate a serious commitment in the form of funding and other resources because mangers, worried that the innovation will weaken their position in the marketplace, will suspend traditional investment screening criteria" (Clark and Bower, 2002: 4). Once the initial resource allocation is secured and the innovation acknowledged as being important, it is then in a second step, time to reframe the innovation from a threat to an opportunity. This is typically the time when a new business model needs to be created and the identification of the demand for the radical innovation should be evaluated. According to Clark and Bower, it is important to reframe the innovation because now managers need to find new and unique applications associated with the innovation. This is best done under the assumption of dealing with an opportunity. If at this moment in time the innovation were still perceived as a threat, managers would react rigidly and apply old models and approaches because comparison and benchmarking to existing innovations has been done.

Strategic options

Floyd (1996: 15ff) and Strebel (1992: 203ff) both handle the problem of uncertainty inherent to the anticipation of discontinuities in the context of creating strategic options. Bearing in mind that forecasting is imprecise and that detected discontinuities might influence the business in a way other than was predicted, it is important to think in options. This means first, that besides investing effort in more than only one of the predicted discontinuities it is recommended and, second, to maintain business as usual. Investing in more than one emerging technology is what Floyd calls "placing side bets." It is the approach of building up competencies and learning what effect specific technologies might have on the existing and future business of the company. This procedure allows a company to react quickly once one of the technologies has demonstrated advantages over an existing one.

The required reaction is twofold: first, the company must change internally from one technology to another as a basis for its products and services and, second, it has to bring this new product and services to the market. There is usually not a lot of time for both of these activities. Experience shows that once a technology is accepted and acknowledged as a challenger in the market, 80 percent of the market can switch within five year, provided that there are no regulatory barriers (Floyd, 1996: 17).

Strebel (1992: 206ff) describes in detail the three steps of building up and exploiting strategic options. They are summarized here:

Preparing the options: Sourcing new competencies, for example techno-logical competencies, that are regarded as crucial for the future. This step is a pre-selection of technologies that have been found to be interesting for the future. They emerge from intelligence work, from continuous improvement, innovation, and from traditional R&D.

Selecting the option: Choosing options to explore potential payoff or promising competencies in the form of new products and processes. Most valuable are considered those options that provide sustainable leverage by projecting existing organizational capabilities and functional competencies into new areas. This advantage can be sustained if the achieved leverage is hard to imitate or buy by competitors and stable against substitutions.

Timing the options: Committing resources fully to the exploitation of the new products and processes. Timing is one of most difficult bets in manag-ing strategic options. A premature commitment risks the initiated projects to run out of resources too early while a late commitment means signifi-cant loss of market share in the emerging market. The best clue for timing an option's commitment is a good environmental analysis that constantly monitors industry forces.

Strategies

Suarez und Utterback (1995) analyze the relationship between technology, firm strategy, industry structure, and the competitiveness of firms in an industry in six different industries. Findings from this analysis have impor-tant implications for companies entering an industry. They suggest that strategies to enter an industry are most successful in the period of pre-dom-inant design. The probability of failure for companies entering a number of years before a dominant design is established in an industry, was clearly lower than when entering in the post-dominant design period (Suarez and Utterback, 1995: 428). The authors explain the success of the early entering companies by the time they have had to learn and experiment with the new products during a period when demand changes were rapid.

Christensen and Raynor (2003) describe the need to master two funda-mentally different strategy processes simultaneously when coping with dis-continuous technologies. This twofold approach is based on the concept of strategy making involving a coexistence of a deliberate and emergent strategy process. This approach was originally described by Mintzberg and Waters (1985). The deliberate strategy aims to organize actions in an organization. It represents a top-down, forward directed strategic approach that fixes goals and coordinates their implementation within the collective interest and intention of the company. The emergent strategy represents a bottom-up approach that emerges from within the working organization doing its daily job. The nature of emerging strategy is more tactical and operational than strategic. Together both approaches eventually determine how a company is going to act strategically, which is finally the realized strategy.

Christensen and Raynor (2003) suggest that success in managing discontinuous technologies depends on the right interplay between both kinds of strategy – deliberate and emergent. Depending on the stage in the lifecycle of a discontinuous technology, one specific approach might lead to success while the other to failure.

At the emerging stage of a discontinuous technology, a deliberate strategy is the approach to avoid. Companies that try to manage their organization according to deliberate strategies in this stage will eventually fail. All the resources spent to find and implement the right strategy will be in vain as the right strategy cannot be known at this stage. In the early stage of a nascent technology, it is the emerging strategy approach that applies best, as the whole dynamics around the technology is discovery driven.

However, once the market for the technology and its applications become clear, companies should change their approach to a deliberate strategy. This change is critical because in this stage, companies have to focus their resources on one common goal, enter, and position itself in the newly opened market. "The switch from an emergent to a deliberate strategy mode is crucial to success in a corporation's initial disruptive business" (Christensen and Raynor, 2003: 222).

The authors suggest a discovery-driven planning method for the management of disruptive innovations as opposed to deliberate planning for sustaining innovations (see Table 9.3). As a first step in discovery driven planning, doing targeted financial projections is suggested. The purpose of this rather unexpected beginning of planning is explained by the following logic: do not lose time in the "cycling charade" of sketching, revising, and tuning assumptions on financial targets, if it is known anyway how good the numbers must look in order to win funding. Instead, first make the targeted financial projections then, in a second step, work out which and how the assumptions can be proven to come true in order for the project to reach the numbers that are expected. In a third and fourth step, implementation is spurred on in order to realize implementation.

Floyd (1996) and Strebel (1992) describe strategies or alternative procedures for those companies that face a discontinuity as latecomers. Floyd

Table 9.3   A discovery driven method for managing the emergent strategy process

Sustaining innovations: deliberate planning	disruptive innovations: discovery driven planning
1. Make assumptions about the future.	1. Make the targeted financial projections.
2. Definea strategy based on those assumptions, and build financial projects based on the strategy.	2. Determine what assumptions must prove true in order for these projections to materialize.
3. Make decisions to invest based on those financial projections.	3. Implement a plan to learn – to test whether the critical assumptions are reasonable.
4. Implement the strategy in order to achieve the projected financial results.	4. Invest to implement the strategy.

Source: Christensen and Raynor, 2003: 430

proposed a choice of alternative procedures for actions rather than real strategies, as they do not relate the company environment to the choice to take. Suggested are five different alternatives (Floyd, 1996: 18):

"Exit," suggests that once a discontinuity has happened it might be worthwhile to consider abandoning the present business and allowing the substitution to happen. This is especially the case when the new technology has by wide margin already become a key selling criteria satisfying costumer needs better that the existing technology.

"Defend the established technology" considers that substitutions do not always prevail and that incumbent technology often can liberate hidden potential. This latter phenomenon is known as "the sailing ship effect" (see also Section 4.1.0. – Life cycle). A further way to defend your own business is to acquire the substitution technology and to bury it. However in most cases this alternative will not work, as driving forces behind a technology that has already proven to represent a threat to existing ones are often very strong.

"Build," better known as the strategic term "make." It is the alternative for a company to develop the substitution technology itself and to catch up with the technological and market knowledge of its competitors. This is often difficult as it involves a great capital investment, time pressure, and the condition to overcome set market barriers.

"Buy" and use the technology is an alternative that includes the possibilities of licensing, acquiring a company, or purchasing components and equipment. Buying is especially advantageous if the technology is not a core technology and not strategically critical as it is probably the fastest way to access a new technology. Its disadvantage however is that the knowledge underlying the technology is hard to transfer to the existing organization so that its development for strategic use for example as a basis for a new core competence will be difficult.

Strebel (1992: 107) goes a step further than to merely describe alternative choices to managing discontinuities as a latecomer. He suggests a catalog of strategies that can be used according to the environmental circumstances that drive the change in the competitive environment of the company. These strategies, called an intervention path include all the efforts to close the deficiencies in functional skills and organizational behavior needed for dealing successfully with the forces of change that could culminate in a discontinuity. The environmental circumstances are described as combinations of forces of change and resistance (see also "Environmental Analysis" in this section) in the company environment. Figure 9.6 displays management intervention paths to choose as a result of the interplay between company external forces of change and resistance. For each intervention path a scope and a pace of change are suggested as well as the approach to close the competence gap in the pursuit of catching up as a latecomer. Scope of change describes how much the organization is involved in the process of acquiring new competence and the pace of change describes the time needed to do so. Strebel (1995: 113ff) depicts the four intervention paths as follows:

Intervention path 1: Resistance. The resistance path assumes that the company can resist or avoid the discontinuity by working on its environment to create more stable conditions where change forces are weakened.

Intervention path 2: Revitalization. Revitalization is appropriate when there is resistance that is still open to change. The goal of this intervention path is to try to lower resistance and to open the way for change forces. Doing so converts the status quo into strong change forces. The scope of change affects the whole organization.

Intervention path 3: Renewal. Renewal is applied when resistance that is open to change must be adapted to a strong but declining change force. Under these conditions, reducing the resistance usually dampens the change force. The scope of the change is limited to parts of the company and the pace is sporadic.

Intervention path 4: Restructuring. Restructuring is appropriate when a strong and growing change force confronts strong resistance that is closed to change. On this path the organization is given a sharp shock to adapt it to the environment. The scope of the change is highly focused, typically on organizational "hardware" such as strategy, structure, and systems.

Strategy implantation

Strategy implementation in the context of discontinuous technologies is described in great detail by Clark and Bower (2002: 6ff). They suggest an

Interplay between forces of change and resistance	Management intervention path	Scope of change	Pace of change process	Approach to closing competence gap
1 Change force strong but declining; resistance closed to change	Resistance	No internal change	Depends on ability to contain change force	No competence gap
2 Change force strong and growing; resistance open to change	Revitalization	Ongoing change throughout the organization	Slow continuous adaption	Long-term investment in organizational learning
3 Change force strong and growing; resistance closed to change	Renewal	Change limited to parts of the organization	Periodic stepwise change	Incremental investment over intermediate period
4 Change force strong and growing; resistance closed to change	Restructuring	Intense change on a few dimensions	Sudden change jumps	Focused investment over a short period

Figure 9.6 Intervention path characteristics

Source: Strebel, 1992: 112

approach derived from insights of corporate venture research. Out of this research, five major points to follow when implementing a strategy with discontinuous technology are:

Separate for better performance: Research findings prove that independent organizations perform better in the implementation projects of a discontinuous technology than projects conducted in the regular parent company. Organizations separated from the core business are usually more innovative and score higher market-penetration rates.

Fund in stages: According to venture capitalists, when financing start-ups in stages a company should also fund its discontinuous innovation projects. Capital is deployed only as the business model takes shape and the market understanding grows. Such an approach demands the project team to seriously perform in order to receive the next round of funding. For the parent company it helps to limits the risk of its commitment to the project as it has the power to stop the project more rapidly when things are not turning out as expected.

Cultivate outside perspectives: The creation of an independent organization is the first right step in providing the initial position for a successful implementation of a strategic discontinuous technology project. However, it does not guarantee that the organization will be managed as an independent one. Crucial for this to happen, is the choice of people selected to carry out the project. Relying exclusively on people from the core organization might thus be the wrong approach as their work processes, decision-making pattern, and focus might be too strongly attached to the core business of the parent organization. Thus, it is suggested to include managers with an outside perspective and a certain distance from the core organization in the project team.

Appoint an active integrator: In order to insure that the independent organization does not shift too much from the interests and positions of the parent company, an active integrator is needed. The person in this position needs to have a great deal of credibility in both organizations so that he can act as a mediator between the two of them. The role of this person is especially important for synchronizing the interests and motivations of both organizations in terms of resources.

Modularize integration: When creating a new organization in order to explore a discontinuous technology there is often the tendency of the parent organization to overemphasize the potential for integration and synergy. Overemphasizing these two aspects might limit the new organization right from the start to fully investigate all of its possibilities as discontinuous technologies typically emerge in new markets and for customers with initially low margins. As time passes and the technology develops, its benefits should then be converged with the core business.

Many aspects of strategy implementation with discontinuous technologies that are covered by the above five points are discussed in a similar way by the other authors. For example many agree with Clark and Bowers proposition that the creation of a separate organization is absolutely critical to the success of a strategic discontinuous technology project. Kusunoki (1997: 381) goes even further than mere organizational separation of discontinuous type technology projects from improvement type projects. His findings suggest that different problem solving approaches are appropriate within these organizations depending on the type of project. In improvement types of projects that rely on evolutionary pattern of technology a product problem solving approach should be applied. However when targeting radical innovation through a discontinuous technology type of project, a technology problem solving approach should be applied. In a nutshell, Kusunoki proposes that organizations carefully separate product development from technology development in order to generate radical innovation. Without contradicting these approaches Tidd (1995: 308ff) points out that the development of complex product systems is likely to require managing across traditional product division boundaries and inside the company. Outside the company strong inter-firm networks are supportive as they can offset missing internal competencies.

Strebel (1992: 143ff) describes different management styles of implementing radical change according to the interplay of forces of change and resistance (see paragraph "Environmental Analysis" in this section) found in the organization and according to the chosen strategic intervention path (see Section 5, Strategies in this chapter). He suggests that especially the forces of resistance determine to which extent the employees, middle management or top management should be involved in the change process. According to their involvement and depending on the pace of change he determines four general management styles: committee style, cultural style, collaborative

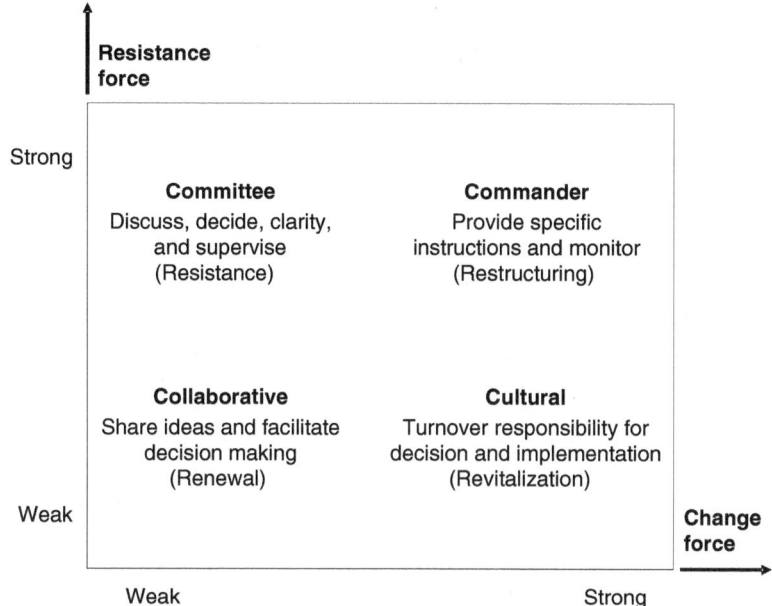

Figure 9.7 Implementation styles
Source: Strebel, 1992: 146.

style, and commander style. Figure 9.7 shows when these styles of management apply and what they mean.

The general perspective over the whole process is delivered in research conducted by Song and Montoya-Weiss (Song and Montoya-Weiss, 1998). This perspective provides empirical support for the notion that different types of projects require different types of management. In a study analyzing 163 radical innovation projects and 169 incremental innovation projects it is shown that radical new product development benefits from careful strategic planning while incremental product development is rather hindered by excessive efforts in strategic planning.

Market perspective

At the industry level, focused research cluster market commercialization issues seem to be less popular only one analysis was found covering this area of research. It is the analysis from Ehrnberg and Sjöberg (1995) looking at the relationship between technological discontinuity and changes in the market shares of companies. Testing a theoretical framework, modeling the variables affecting this relationship in three different industries they find that the faster the diffusion of a discontinuous technology, the greater the probability that early movers will gain initial advantage. Furthermore, the faster the diffusion is, the greater the possibilities that early movers can build sustainable, volume-related entry, and mobility barriers.

The implication of these findings from a strategic management perspective of a company is that "the time available for detecting the need to change and to act is limited by the market growth of the new product" (Ehrnberg and Sjöberg, 1995: 93).

Industry level focused research: seven management recommendations

All authors presented above describe strategic approaches for the management of discontinuous technologies and radical innovation.

Five of the authors give recommendations for organizational structures. All of them suggest a separation of the discontinuous technology project from the core business and its development in an independent organization. In this context Kusunoki even shows evidence that a separation of technology development from product development in separately managed organizations are most helpful in order to successful promote radical innovation.

In its essence the findings from industry focused research on strategic management with discontinuous technologies and radical innovation can be summarized in the following seven management recommendations:

Create an organization separated from the core business for the management of discontinuous technologies and radical innovation.

Differentiate the management of technology development and the management of product development.

Manage early endeavors with discontinuous technology and radical innovation as a portfolio of strategic options with the aim of long-term competence learning.

Build and foster firm internal and cross firm boundary spanning networks to (1) eliminate uncertainty, and (2) influence emerging industry standards.

Let the organization perceive a discontinuous technology as a threat. Let the organization implement a discontinuous technology as an opportunity.

Enter competition with a discontinuous technology based product in a predominant design phase to ensure higher success rate in the market.

Master the interplay between both strategies: deliberate and emergent, according to the life cycle of the discontinuous technology.

Company level focused research

Generalized management perspective

The generalized management perspective on research at the company level is structured in the same way as research on the industry level focus (see

Table 9.4 Generalized management perspective of company level focused research and its contribution to the generic strategy development process

Authors	Year	a. Strategic objectives	b. Environmental Analysis	c. Company Analysis	d. Strategic Options	e. Strategies	f. Strategy Implementation
Maidique	1984	■					■
Strebel	1995	■	■				■
Rice	1996				■		■
Rice	1998	■	■		■		■
Leifer	1998	■					■
Veryzer	1998a	■	■	■			■
Markham & Giffin	1998						■
Kessler	1999	■			■		■
Lois	2000						
Rice	2000		■				
Rice	2001	■					
Savioz	2002	Describes an own strategic process					
Salavou and Lioukas	2003	■					
O'Reilly & Tushman	2004						■

Section 4.1.0). Along the generic six steps of the strategy development process as shown in section each author's contribution to a specific step is presented. Table 9.4 shows the overview of authors presented in this section as
well as their contribution to strategy development when managing discontinuous technology and radical innovation.

Strategic objectives

The majority of the authors in this cluster dedicate one focus of their
research activity to the issue of setting strategic goals (see Table 9.4). Most
agree implicitly with Maidique and Hayes' (1982) argument that strategic alignment through strong leadership is essential when the future is
uncertain and changing rapidly. When managing discontinuous technologies and radical innovation, the future typically has such characteristics. In this situation hands-on top management is needed that promotes
an entrepreneurial culture and that creates trust within the organization
for these most often quite visionary projects (Maidique and Patch, 1982:
23; Salavou and Lioukas, 2004: 101ff). However, in order to strategically
manage visionary projects successfully, clearly setting boundaries for the
project, defining its strategic intent, and indicating the business field to
enter are critical tasks to manage during strategic goal setting (Rice et al.,
1998: 54). Besides these tasks that create the premises for a project start
it is often the initiation of the latter that encounters difficulties. Thus,
Rice et al. (2001: 54) suggests a set of questions that, used as a tool checks
the "radicalness" of a project. It includes questions concerning technology, market, and corporate strategy related issues. It helps the technologists that are most often the initiators of radical innovation projects, to
argue in favor of such projects more easily when confronted by middle
management. In many cases the initiation of radical innovation projects
fails as technologists are not familiar with strategic language. They have
a hard time convincing management of the strategic impact of projects.
Although the bottom-up initiation of radical innovation projects carried from a technologist up to management might be well prepared, it
still needs a member of senior management who is creative and sensitive
enough for a project to be started. As Strebel puts it: "top-management
cannot abdicate its entrepreneurial role" (Strebel, 1995: 20). This is especially the case as radical innovation projects take considerably longer to
develop than incremental projects. Without a certain protection from top
management, such projects have no chance to survive strategic project
management meetings over years and years (Leifer, 1998: 134).

Kessler and Chakrabarti (1999: 239) found that the long development
time of radical innovation projects can be shortened considerably by setting clear goals for such projects, in contrast to incremental innovation
projects, where the reverse was observed. These and other fundamental
differences in the management of radical and incremental innovation are

Table 9.5   Ambidextrous leadership

Alignment of:	Exploitative business	Explorative business
Strategic intent	Cost, profit	Innovation, growth
Critical tasks	Operations, efficiency, incremental innovation	Adaptability, new products, breakthrough innovation
Competencies	Operational	Entrepreneurial
Structures	Formal, mechanistic	Adaptive, loose
Controls, rewards	Margins, productivity	Milestones, growth
Culture	Efficiency, low risk, quality, customers	Risk taking, speed, flexibility, experimentation
Leadership role	Authoritative, top down	Visionary, involved

Source: O'Reilly and Tushman, 2004: 80

the initial position for O'Reilly and Tushman (2004: 80) that make a claim for an ambidextrous leadership. This approach is based on the heritage of the research conducted since the late 1980s by a group of researchers from Stanford and Harvard University. This leadership focuses its strategic goals on two types of businesses – those concerned with exploiting existing capabilities for profit and those concerned with exploring new opportunities for growth. Table 9.5 summarizes the essence of this leadership approach from different management aspects that need to be directed despite the twofold focus.

Environmental analysis

Veryzer (1998a: 319) suggests that before going into a detailed environmental and market analysis, it is especially important in the case of discontinuous technology and radical innovation to formulate beforehand initial applications of the technology. The explanation of this procedure is due to the nature of such technologies and innovations; they tend to be further removed from the market and are more technology rather than market driven. Thus, it is good to have a target application before going into market research for opportunity recognition. According to Rice et al. (1998: 57) and Strebel (1995: 19) this opportunity recognition is best done by first line or front line managers – not by senior managers. In his study it was first line managers that were most successful in initially identifying future market opportunities. These are people with entrepreneurial character, who are leading improvement teams. Aided by an informal network operating between R&D and business-units and between R&D and outside constituents like customers, suppliers, and governmental agencies they can eliminate uncertainty inherent to radical innovation and discontinuous technology can best be eliminated (Rice, 1996: 531). Despite their good performance in recognizing business opportunities, front-line managers did not always get the attention necessary to push the opportunity through in the organization.

Once a market opportunity is found and an application is set, companies need a corporate intelligence system for collecting and analyzing data from the relevant environment with regard to the planned discontinuous technology or the radical innovation to commercialize (Strebel, 1995: 15). Such intelligence systems should include formal and informal systems, such as "environmental scanning systems, benchmarking systems, information systems for gathering and distributing information and insight, and an integrated intelligence consciousness that involve all executives in learning from the past success and failure to project how alternative decisions might play out in different environments..." (Strebel, 1995: 15).

Company analysis

For the step describing company analysis only one contribution could be found in literature: Veryzer (1998a) suggests that measures, which evaluate the impact of a discontinuous technology or a radical innovation are important for the company and this evaluation fundamentally differs from that of incremental projects.

Veryzer (1998a) however could not find in any of the analyzed companies a deliberate evaluation process for the purpose of radical innovation and discontinuous technologies. Rather he found that such evaluations are part of the normal evaluation process of incremental innovation projects or treated in an "ad hoc fashion" (Veryzer, 1998a: 55). In some cases when traditional evaluation criteria and methods were used they were generally not recognized as relevant to the decision.

Strategic options

Rice et al (1998) do not explicitly refer to strategic options in this article however the approach described can be interpreted as an options building approach. They call for an assessment of discontinuous technologies and radical innovation that is "market learning more that market evaluation" (Rice et al., 1998: 56). They point out that it is not primarily to assess the impact on sales but rather how the market records the "value-in-use." He argues that, at a stage where prototypes are still quite primitive, getting an idea of what the technology could look like in the future is mainly a stage of learning. It is learning about whether and how the market will receive the technology. Thus, with this learning several strategic options can be built-up. These options in learning are predecessors to future company competencies. They still leave the choice of further investment or no further investment depending on the development of the overall uncertainty inherent in the discontinuous technology. This choice is reflected by the name "option." Options are one very effective way of managing through times of uncertainty (Peter and O'Connor, 2002: 616).

Strebel (1995: 19) describes organizational issues within the context of building up and managing strategic options. As already described in the previous section front-line managers are best suited for identifying opportunities that can be explored as strategic options. Then middle managers should manage the portfolio of options generated by the frontline managers. Middle management has the task of bringing together the vision of top management and the uncertain reality of the market place by deciding with the front-line managers in which options to invest.

Strategies

There are no contributions of the authors presented in this chapter for the "strategy" step.

Strategy implementation

In strategy implementation most authors seem to agree that radical innovation and incremental innovation need different implementation approaches. Maidique and Patch (1982: 26ff) explain this necessity by the conflict triggered through the different kinds of innovations. The conflict is based on contradictory goals that have to be managed simultaneously: on the one hand, continuity sought by incremental innovations, on the other hand, chaos triggered by radical innovation. They speak of a paradox of continuity and chaos in managing stability and change. In order to avoid this management paradox a separated management of radical innovation and discontinuous technologies on the one side, and incremental innovation and continuously evolving technologies on the other side is needed. In this context Strebel (1995: 18) speaks of the necessity to manage directed and spontaneous innovation. By directed innovation he refers to systematized capabilities of mass production and continual improvement

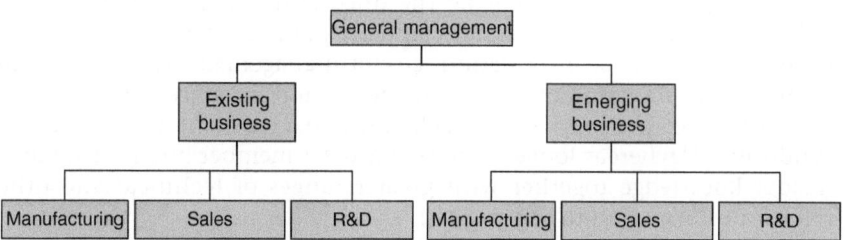

Figure 9.8 Ambidextrous organization
Source: O'Reilly and Tushman, 2004: 79

while by spontaneous innovation he means the breakthroughs in technology that lead to discontinuities and radical innovation. In this context Maidique and Patch (1982: 27) refer to management that extends the past and management that breaks with it. O'Reilly and Tushman (2004: 79) suggest a separated organization according to the type of business calling it an ambidextrous organization: "established project teams that are structurally independent units. Each having its own process, structures and cultures, but are integrated into the existing management hierarchy." Figure 9.8 shows the core idea of an ambidextrous organization.

Within such an ambidextrous organization, where incremental innovation projects are separated from radical innovation projects, the latter is best supported if champions[9] are involved (Leifer, 1998: 134). Champions are people that "are entrepreneurial in accessing resources to accomplish a mission, and are action oriented and focused" (O'Connor and Veryzer, 2001: 239).

Kessler and Chakrabarti (1999: 241) show that development speed in radical innovation projects is dependent upon its staffing. Their findings suggest that a greater number of champions than usual deployed in incremental innovation projects speed up the development of radical innovation projects. Further they show a relationship between the speed of radical innovation development and the choice of the project leader and project members. For a radical innovation project, speed increases the higher the project manger is positioned and the shorter his tenure. For project members however, longer tenure most speeds up the development of the project. As an origin to this observation, Kessler and Chakrabarti (1999: 241ff) make the following assumptions: the greater number of champions needed to navigate a radical innovation project through the organization might be due to the broader focus of such projects affecting a wider rage of different units in an organization. Thus, such projects require more people for an effective information exchange that is spread throughout the whole company instead of just being concentrated in one location. Concerning the higher position of the project leader, there is the assumption that it might be the uncertain character involved in a radical project that needs greater company political influence in order to promote and control information, co-opt management, and build coalitions. One explanation of the advantage of shorter tenure of project leaders is that they are supposed to feel a lower degree of "not invented here syndrome,"[10] whereas longer tenure of project member might bring more insider knowledge together with greater ranges of technical and other relevant information to the project.

Main targets to achieve in radical innovation projects should be directed towards early prototyping, conducting pilot processes, and carrying out test marketing (Strebel, 1995: 13; Veryzer, 1998a: 319). The organizational elements for such goals to achieve are incentive structures that reward people for identifying opportunities and encouraging them for calculated risk taking (Maidique and Patch, 1982: 24).

While all authors describe the necessity for radical innovation to be separated from the core business Rice et al. (1996: 587) examine how to transfer (separated) discontinuous innovation projects to operational status once technological and market challenges are know. They suggest seven key points for transition management: (1) create a transition team, (2) assess transition readiness, (3) develop a detailed transition plan, (4) identify transition senior management, (5) establish a transition team oversight board, (6) provide transition funding and commitment, and (7) lay the groundwork for a big market.

A prescriptive strategy development process

Savioz et al. (2002) are the only group of authors modeling a prescriptive process model for the strategic management of discontinuous technology and radical innovation. However, their model exclusively concentrates on the "fuzzy front-end" of innovation. The fuzzy front-end of innovation refers to the very early stage of the innovation management process which decides which development projects should or should not be carried out. The process is understood as a gradual process with the eventual target of strategic goal definition. It is considered an open system with input and output. Input is information about technology and market change as well as of all competencies existing in the company. Output is selected projects assigned with resources. The cases conducted in this research work suggest five distinct tasks to be executed in the process. These tasks should be considered as tasks to process stochastically and in

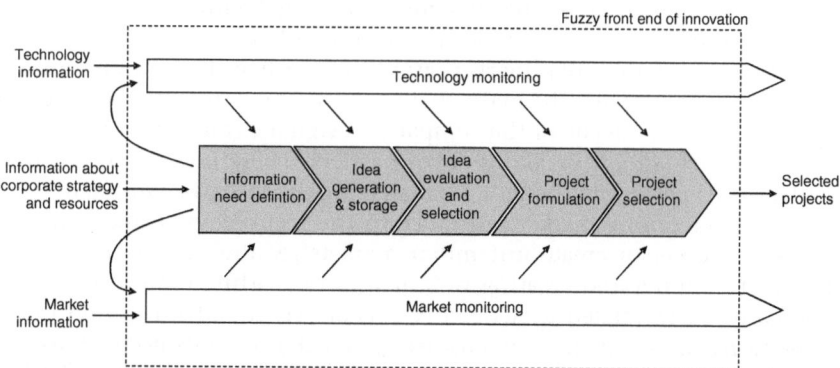

Figure 9.9   System "Fuzzy front end of the radical innovation process"
Source: Savioz et al., 2002: 400

parallel rather than sequentially. Those tasks are: (1) Definition of information need, (2) idea generation and storage, (3) idea evaluation and selection, (4) project formulation, and (5) project selection. A continuous technology and market scanning support these five steps. The process is illustrated in Figure 9.9.

Many of the management aspects named in this article (Savioz et al., 2002) have already been mentioned in the previous sections. Thus, instead of reviewing the whole article, this section will emphasize only two additional new management insights: First, that the above shown process should be a continuously conducted process. It should not be executed as a project for example for the annual update of the strategy. It is rather meant to continuously identify and deliver radical innovation ideas. Second, in order for such a process to work as designed, a well-defined and long-term corporate strategy is needed as an initial position for the process. Only a visionary and far-sighted corporate strategy is able to set guidelines for the definition of information need and to set criteria for evaluation and selection of ideas and projects.

Market commercialization perspective

Almost all authors dealing with market commercialization of discontinuous technologies and radical innovation agree that the challenges to master are different from those dealing with continuous technologies and incremental innovations. Before looking at the solutions described in market commercialization of discontinuous technologies and radical innovation, a brief overview of the main challenges related to this subject is given.

McDermott and O'Connor (2002: 427ff) observe in a longitudinal multidisciplinary study of radical innovation projects the challenges associated with such projects. Their findings are grouped into three high-level strategic themes:

Market scope: discusses the challenges associated with the choice of placing radical innovation in familiar versus unfamiliar markets. Radical innovations that enter familiar markets aim to strengthen the firm's position in this particular market. The main challenges that such innovations encounter – within and outside of the company – originate from either encountering resistance and/or breaking down barriers. These include (1) ensuring delivery of a perceptible benefit, (2) managing the threat of cannibalization, and (3) overcoming market resistance to the technology. Radical innovations that enter or create unfamiliar markets require in proactive investment in building and creating new areas both within and outside of the company. Tasks related to achieving this are (1) the identification of the operational home of the new business given that there is no obvious link to existing units, and (2) the creation of an effective business model that include the potential the innovation offers.

Competence management: discusses strategic challenges related to the extension of existing company competencies into new domains. The challenges related to the extension of competencies imply primarily the management of the risks associated with destruction, enhancement or stretching of competencies.

People issues: discusses the consequences of people and teams trying to move radical innovation projects forward in an organization that is not prepared to support the uncertainty inherent to these projects.

As overall management implications from these challenges, the authors suggest that the role of project management in dealing with radical innovation is less important than the management of uncertainties. Uncertainties of technological, market, and organizational nature have to be managed. Furthermore, the beneficial role of informal networks was emphasized as well as the role of people willing to identify and promote high risk and high potential projects within the firm. The importance of reintegration plans for radical innovation that is developed outside an established core business was also mentioned.

Walsh et al. (2002) examine which type of firms – established large firms or small new ones – are best at taking the challenges of commercializing radical innovation. They empirically show that established firms rarely commercialize disruptive technology, and only when pulled by the market (2002: 349). New firms commercialize primarily discontinuous technologies and perform better in it than established companies. Their time to market is about one fourth of that of the established firms. It is this time advantage as well as their flexibility in marketing that is the root of their success. Marketing flexibility means the freedom new firms enjoy by not having existing customers that influence the innovation efforts of a company into a direction preventing them from commercializing radical innovation.

The main difference in commercializing discontinuous technology and radical innovation compared to continuous technology and incremental innovation is related to the objective of the innovation. According to most authors (Lynn et al., 1996; O'Connor, 1998; O'Connor and Veryzer, 2001; Kassicieh et al., 2002; McDermott and O'Connor, 2002) presented in this section the primary objective established companies should have when entering discontinuous technology development is learning new competencies. Thus, O'Connor (1998: 162) points out that the assessment of market potential, size, and growth is no issue during the early stage of a radical innovation endeavor. It is rather the assessment of what competencies the company can build up in order to position itself well strategically in the long-term (Kassicieh et al., 2002: 385). In most of the cases, the uncertainty evolved with discontinuous technologies and radical innovation prevents traditional quantitative market assessments. In such cases mathematical models of market diffusion such as those presented by Linton

(2002: 371) might be hard to apply as the data to feed the models are not readily available.

Often it is even unclear which market applications a technology should be best used, as even customers have a hard time evaluating the benefit of such applications. In his study, Veryzer (1998: 143) analyzes key factors affecting customer evaluation of discontinuous new technology application. He shows that six main factors were at the origin of the difficulties customer have evaluating new products: (1) lack of familiarity with the product, (2) irrationality in evaluation, (3) user-product interaction problems, (4) uncertainty and risk concerning the benefit of the product, (5) accordance, referring to the compatibility of the new product with the customer's life or business situation as well as the amount of accommodation or adjustment required by the innovation, and (6) aesthetics influencing customer reaction to discontinuous products in the same way it influences their reaction to continuous products.

In order to bring clarity to an uncertain market situation where neither quantitative assessments nor customer insight is available O'Connor and Veryzer (2001: 232) suggest a process of market visioning. Analogous to the innovation management step "imagining" proposed by Jolly (1997: 3), visioning aims to mentally relate a discontinuous technology to a potentially attractive market opportunity. According to the findings of O'Connor and Veryzer (2001: 234ff) there are a number of drivers that initiate and sustain the vision, such as: the participation of senior management as a promoter of a company wide focus for innovation and as a communicator of goals, or scientists with an understanding for business that could act as opportunity recognizers. The process is best conducted when champions, "implementers" and "ruminators" are involved. Champions have already been mentioned in this thesis (for example, see the section Strategy Implementation), implementers are people that enjoy participating in projects and have the potential to trigger major changes in the organization, ruminators are contemplative, and experienced people who spend their time thinking about the future and that have the ability to break with the bounded company view.

Once the link between a discontinuous technology and a potentially attractive market application is found, Lynn et al. (1996) suggest to proceed according to the "probe and learn process." "Probe" means experimenting with prototypes and introducing early versions of products into the market. However, probing only makes sense if it is done with a higher strategic goal. The goal is typically to "learn" in order to enter into or to build up a market that is new for the company. It has nothing in common with a trial and error approach as it is done within a strategic context with defined corporate goals. It can rather be compared with an online and stepwise learning procedure as the feedback generated from a first "probe" or attempt can instantly be used for a second one. Doing so, the discontinuous technology is usually first commercialized in applications other than the finally targeted one and in a market still familiar to the company. The idea behind this procedure is to apply the technology in products that do not yet require a performance as high as the final one and to learn how the market will react. Instead of merely developing the technology's performance until it has reached the final level required in order to enter an unfamiliar market, it is better to commercialize it first in familiar markets with lower performance requirements as early as possible in order to learn about its acceptance and technological possibilities step by step and then to approach an unfamiliar market. This way the technology is commercialized as soon as its performance level fits to a first application. Besides reducing the total risk of a new market entry in manageable risk entities, this procedure is also favorable to the approach of "fast failure." Fast failure describes the approach of companies that try to stop projects immediately as soon as their failures become visible. The "probe and learn" approach allows failure to become visible much faster than if a technology is held back from its eventual customers until it reaches the level of performance necessary to enter its finally targeted market.

The "probe and learn" process however bears one disadvantage not to be underestimated: it is the risk of commercializing a technology in a stage that is too early to be already enough well developed or known. Such a technology might reveal side effects or have consequences unpredictable in an early commercialization. At a latter moment in time these undesired effects could have been relieved and patched. Thus, the probe and learn process bears the inherent risk of scattering a potential future market before it has even been created.

Beside the probe and learn process Wood and Brown (1998: 182) describe a process of commercializing nascent discontinuous technologies in a three step process:

Approbation: nascent technologies are monitored, assessed, and captured based on the anticipated needs of development. As premises for the approbation step, the authors require ideally a small loosely structured research organization where the coordination between the research organization and the development organization is well managed. This coordination should help focus research on the problems most important to development; it keeps development sensitive to emerging technologies and eventually facilitates the knowledge transfer from research to development. In order to facilitate the transfer some methods are suggested such as: research exhibitions, internal conferences, and career paths spanning research, development and operations.

Implementation: a technology is refined to the point where it can be produced and controlled, at which point the technology is incorporated into development projects. At this stage, an effective transfer from research to development is required in order to get the technology to a maturity level where it is reproducible, testable, and documented. To ease technology transfer to development, it is suggested to create explicit project teams charged with achieving implementation readiness. Such teams can be transferred along with the technology from the research organization to the development organization.

Manufacture: large-scale production systems for products using the new technology are developed and refined in response to changing volume and cost requirements. For this stage an internal production engineering function is suggested especially for the commercialization of nascent technology. Such a function is recommended as a special set of skills, different from tradition skills in research, development or operations are needed in order to design, construct, and refine new production systems.

Collaboration and acquisition perspective

An increasing number of established firms consider collaborations or acquisitions as a reaction to a technological discontinuity (Lambe and Spekman, 1997: 107). The popularity of this reaction can be explained by two main requirements that emerge when a discontinuity occurs: first the need for a rapid reaction and second the need to eliminate the triggered uncertainty in the industry. Collaborations or acquisitions are one

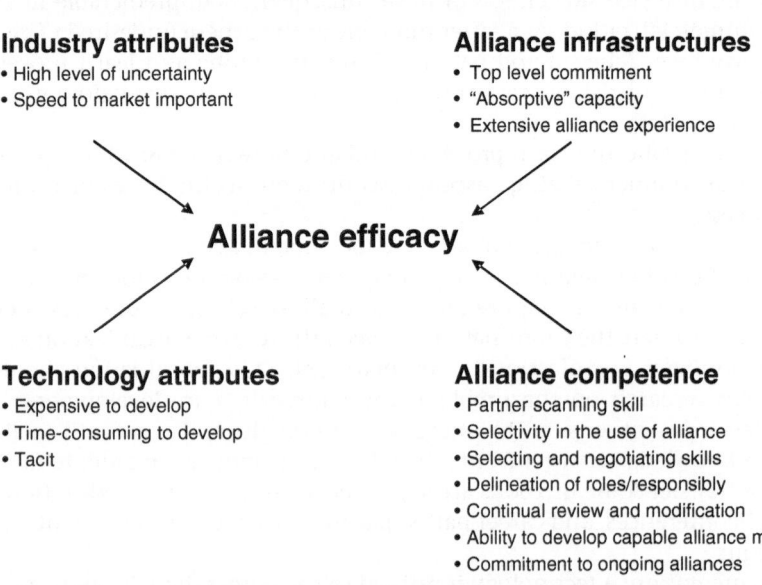

Industry attributes
- High level of uncertainty
- Speed to market important

Alliance infrastructures
- Top level commitment
- "Absorptive" capacity
- Extensive alliance experience

Alliance efficacy

Technology attributes
- Expensive to develop
- Time-consuming to develop
- Tacit

Alliance competence
- Partner scanning skill
- Selectivity in the use of alliance
- Selecting and negotiating skills
- Delineation of roles/responsibly
- Continual review and modification
- Ability to develop capable alliance managers
- Commitment to ongoing alliances

Figure 9.10   Factors that contribute to the efficacy of technology-sourcing alliances
Source: Lambe and Spekman, 1997: 112

of fastest ways to adopt a discontinuous technology and the joint works they promote between companies are an effective way to respond to general uncertainty. The most common form of collaboration to respond to a discontinuity is a strategic alliance (Lambe and Spekman, 1997: 108). Such strategic alliances are especially attractive as they require much lower overall investment and pose considerably less risk than potentially failed acquisitions or mergers.[11] Lambe and Spekman (1997: 111) describe four premises to enhance the efficacy of alliances to source technology in the face of a discontinuity (see Figure 9.10): (1) the creation of an infrastructure as a part of corporate technology strategy that supports both strategic partnering and the development of an alliance competence, (2) an alliance competence, which includes extensive partner scanning capabilities and the ability to selectively use alliances to source technology, (3) technology acquired through alliances should be tacit, expensive, and time-consuming to develop, and (4) managers should anticipate and plan for uncertainty effects of technology irrelevance and diminish alliance motivations.

The assessment of potential alliance partners is also analyzed by Hruby et al. (2000). They suggest a five step procedure for selecting partners (see Figure 9.11). Unfortunately the procedure has only been validated in one case and might not be suitable as generic procedure.

Once these premises are set and a suitable alliance partner is identified and selected, the timing of the collaboration with regard to the life cycle of the discontinuity should be carefully chosen. Peters (1996: 473) points out that the starting time of most such alliances is only after the

Determine economic
viability of firm
↓
Examine resources
available to partners
↓
Focus on firms with
high potential
↓
Examine strategic
intent of firms
↓
Identify areas of use
within industries and
within firms

Figure 9.11　Steps in assessing partners in technology commercialization
Source: Hruby et al., 2000: 339

established company has initially developed its key technology-product concept itself. He describes the benefits of alliances as multiple: alliances bring credibility as well as reduced risk to an established firm's discontinuous innovation project. Alliances leverage a firm's technology capability to extract the highest possible value from the technology investment and it can enable entrance into new markets. Furthermore, the networking activities as a consequence of the alliance appears to be important for building up corporate commitment to commercialization (Peters, 1996: 473).

Besides collaboration between companies Peters (2000) points out that industry university collaborations are very successful in developing discontinuous technologies. In his study he empirically shows that as a focus of activity in radical innovation, universities are the most often mentioned partners outside of the firm. However, the role of universities is most often discussed as an afterthought. He develops a hypothesis suggesting opportunities for a more effective exploitation of academic research in radical innovation. Unfortunately these hypotheses remain unverified.

Company level focused research: seven management recommendations

Similarly to research conducted on industry level, most authors focusing research on the company level claim that discontinuous technology and radical innovation should be handled differently and separately from continuously evolving technologies and incremental innovation. Thus one major recommendation confirming findings from the industry focused group of researchers is to separate and manage organizations differently according to their innovation focus: incremental innovation or radical innovation. Concretely speaking the recommendation is to take out of the core business a group of highly qualified researchers in order to create a separated group responsible for the realization of a given radical innovation project.

Summarizing the specific findings from the research conducted on company level also leads to seven distinct management recommendations:

Set clear strategic goals to shorten radical innovation projects' durations.

Design imagining and visioning processes to facilitate the link between discontinuous technology and its potential market application.

Involve a high number of champions in projects who are willing to promote high risk and high potential projects.

Assign highly positioned project leaders with short company tenure, and experienced team members with long-term tenure in the company.

Realize early prototypes, process pilots, and market tests, probe and learn within a set strategic context with a long-term competence building perspective.

Collaborate through strategic alliances with industry and academia in order to eliminate uncertainty and influence industry standards, leverage credibility, and reduce risks.

Ensure that front line managers are charged with opportunity recognition to achieve the greatest success.

Conclusion

This article first gave an overview of the common understanding of what scholars call disruptive technologies and radical innovation. It also showed the various research directions those scholars are investigating and presented their management solution approaches. Despite those many valuable insights on how to manage the challenges of discontinuous technologies and radical innovation, none of the authors showed how their solution approaches should be integrated into existing management practices alongside the majority of companies more usual incremental innovation efforts. The research community acknowledges that major challenges of radical innovations not only lie in its own nature but, also to a great extent, have conflicting innovation management goals created by the necessity to manage both kinds of innovations – radical and incremental – at the same time (refer to 0).

Best practice cases

A possible selection of practice cases can be added in a second chapter relating to the findings of this article that show how the distinction between radical innovation management and incremental innovation management can be implemented in practice.

BASF: different R&D activities according to different levels of market and technology newness

BASF classifies innovation activities according to different levels of technology and market newness.

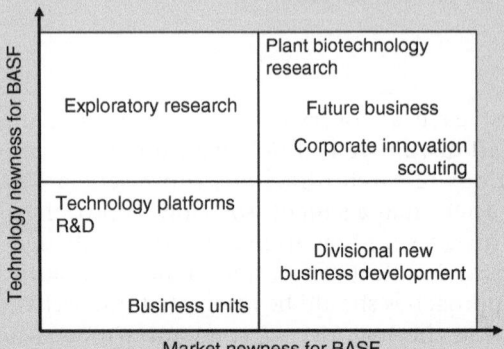

Market newness for BASF

Technology platform research and development refers to R&D activities focusing on technologies andmarkets that are established within BASF. The three technology platforms in the fields of specialty chemicals, chemicals and engineering, and polymers, perform research and development activities in order to strengthen and extend the core competencies of the company.

Exploratory research is a strategic instrument to secure BASF's technological and methodological competence in the long-term. With that, the potentials from scientific and technical progress for the existing portfolio as well as for new attractive business areas are exploited. Basic research activities are left up to universities to which BASF's research units have good contacts. Most of the BASF exploratory research is carried out in Germany.

New business development is conducted in the operating divisions of BASF. It aims to identify new markets outside the current businesses that are based on existing and already mastered technologies. The focus lies on the development of new partnerships with customers, start-ups or universities, utilizing the expertise available in-house.

Plant biotechnology research is conducted within BASF Plant science GmbH, a spin-off company of BASF founded in 1998. It focuses on "young" technology in the area of consumer-oriented and value-adding development of plants.

Corporate innovation scouting is conducted within Strategic Planning, a unit with tasks in product area strategy definition, mergers and acquisitions, strategic controlling, competitive intelligence, and macroeconomic analysis. They cooperate with Corporate Research Planning on planning and reporting on corporate financed research as well as division-based R&D. Innovation Scouting is a global platform for the entire BASF group used to identify new cross-divisional business opportunities, evaluate those, and propose an appropriate model of how to address these new opportunities. The understanding of the term "innovation" is not limited to new products, but also to new business models based on existing products from more than one division. So-called "innovation scouts" will trigger and support the idea creation process, work on business opportunities, leverage existing and develop the future product portfolio, as well as create innovative business solutions.

BASF Future business GmbH is a 100% BASF subsidiary with the task to identify and develop new business areas for BASF group.

Figure 9.12   BASF: different R&D approaches according to different levels of market and technology newness

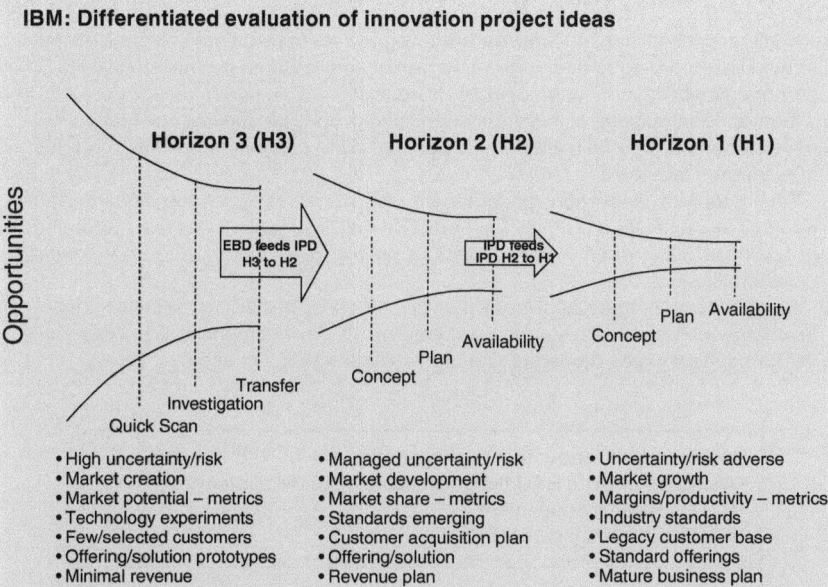

Similarly to the differentiating analysis in the subprocesses "Investigation" and "Evaluation" of the suggested strategic planning model in this research, IBM uses different types of measures to assess different types of innovation project ideas. After having clustered project ideas according to three different horizons (see Best Practice Case 1), IBM uses differentiating priorities according to various horizons. Using an integrated portfolio development (IPD), the ultimate goal is to filter an initially great number of emerging business opportunities (EDO) step by step from horizon three to horizon one and to build up new core businesses managed in H1.

Figure 9.13 IBM: different evaluation of innovation project ideas according to risk and uncertainty

Degussa: different R&D responsibilities according to different innovation management foci

R&D responsibilities at Degussa are assigned to three different organizational structures according to their innovation management focus: business units' R&D, corporate innovation management, and Creavis.

Innovation management in Degussa's business units pursue low risk technology development and enhancement of existing core competencies following the business unit strategies.

The corporate innovation management unit leads research coordination and strategy at Degussa. It diffuses best practice management processes elaborated in business units, directs communication among researchers, and coordinates collaborative projects.

Creavis Technologies and Innovation is run as a unit within Degussa that is in charge of Degussa's strategic research and development and a corporate venturing. It manages moderate and high risk innovation projects.

Business unit innovation management
- Allocation of BU´s R&D budgets according to BU strategy
- R&D portfolio management
- R&D project management

Corporate innovation management
- Corporate guidelines (best practice)
- Allocation of corporate R&D funds according to corporate strategy
- Check and balance process of BU´s technology positions
- Support of BU innovation through coordination of internal/external R&D-networks and R&D-information systems

---

Creavis Technologies and Innovation
- Strategic radar, technology watch
- Idea management
- New business development (outside existing portfolio)

Figure 9.14   Degussa: different R&D responsibilities according to different innovation management foci (Creavis)

ABB: product generation focused R&D structures

ABB basically structures its R&D activities on the divisional as well as on the corporate level according to their two main segments: (1) Automation technologies and (2) Power technologies.

The divisional R&D activities in these two main segments are further structured according to business areas such as automation products, manufacturing automation, etc., in segment automation technologies and power systems, medium-voltage products, high voltage products, etc. in the segment power technologies. Each of theses business areas has a Business Area R&D Manager that focuses and coordinates activities for the development of present and next generation products. They define their own product and systems strategies. Local R&D representations of the Business Area R&D in countries are Business Area Units (BA-U. R&D).

The corporate R&D activities led by the CTO are conducted in ABB's two global labs: one focusing on automation technologies, the other on power technologies. These labs, locally represented in research centers around the world, develop the technologies for the second next and beyond product generations. Those technologies are explored within interdisciplinary research programs that cover areas of emerging technologies.

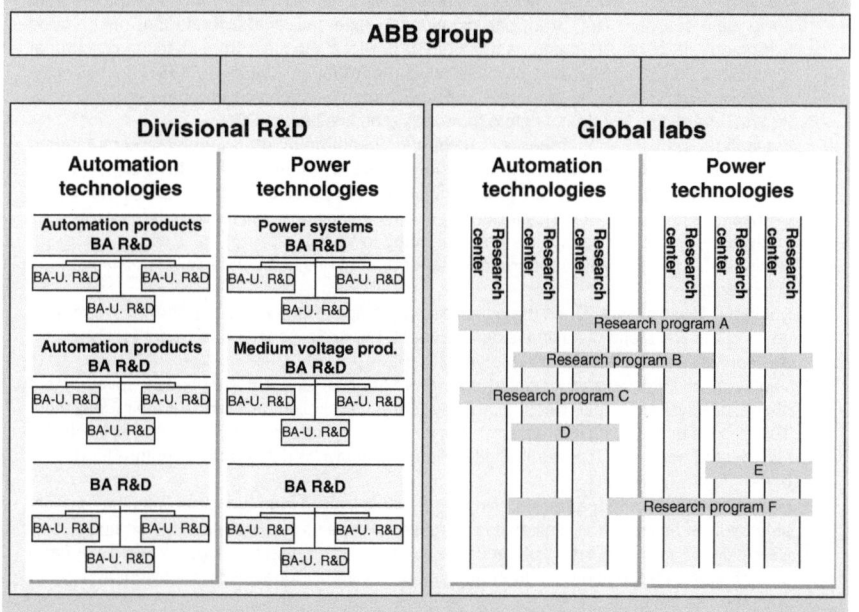

Figure 9.15   ABB: product generation focused R&D structures

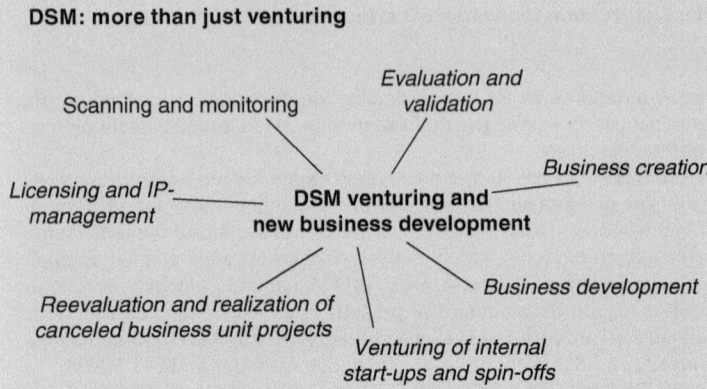

DSM: more than just venturing

DSM's venturing and business (V&BD) development organization is a best practice example showing which variety of functions such an organization can cover in order to foster innovation, especially radical innovation in an established company. It is run similarly to a business unit as an autonomous entity of 100–150 people within the company. The following describes the main activities of this unit.

- Scanning and monitoring: V&BD support intelligence work by gathering information in wider fields than the ones covered by the business units. This information originally comes from three main sources: First, from participations in external venture funds that have young technology driven start-up companies in their portfolio. Second, from all kinds of external technology and business ideas proposed by entrepreneurial people as V&BD is positioned in DSM as a drop-in center for external ideas. Third, by recognizing company internal ideas that are interesting but do not fit into the existing businesses of DSM.
- Evaluation and validation: Ideas that have been captured by the scanning and monitoring routines are evaluated and validated for their business potential by feasibility projects conduced within V&BD. Technology and market feasibility studies are conducted with the goal to check the proof of concept for emerging technologies and novel business ideas. For such analysis business unit specialists are contracted to V&BD.
- Business creation: V&BD supports the elaboration of business plans for ideas with a high business potential.
- Business development: Once the business plan is approved, V&BD support the implementation of the first steps of business development. For this purpose projects are launched, for example for the elaboration of prototypes.
- Venturing of internal start-ups and spin-offs: After a successful business development phase, projects are transferred into an independent organizational form for their realization. This is done by founding internal start-ups for company related opportunities or spin-offs for less related opportunities. Additionally, venturing is involved in external venture funds and external start-ups.
- Licensing and IP-management: Selling internal knowledge from business units that is available for externalization as well as procuring knowledge that is needed in the company.
- Reevaluation and realization of canceled business unit projects: Projects that have been dismantled in the business units are picked up by V&BD for a second reevaluation and a possible realization.

With all these activities, V&BD represents a unit within DSM that embodies technology change. It not only procures and develops radically new technologies and business ideas for the company by scanning, monitoring, assessing, and finally developing them for their deployment in existing and new businesses. It also helps the company to "get rid of" technologies and businesses by externalizing them through licensing and selling them.

Figure 9.16   DSM: more than just venturing

Clariant: idea management and project realization adapted to different levels of risk and newness

Within Clariant's business divisions there are two distinct treatments by which innovation ideas are generally managed and realized in projects.

Technology and business opportunities related to the divisional core business and representing a moderate level of risk are assessed within the business divisions. The realization of such ideas in concrete projects are executed in the corresponding divisions, without any interference from the corporate level or another divisions.

Opportunities formulated by the divisions that are however out of the scope of existing divisional fields and implying a higher degree of risk are forwarded to an R&D Council to be discussed. This council is further provided with innovation opportunities of corporate concern gathered by the technology and innovation management entity on the corporate level.

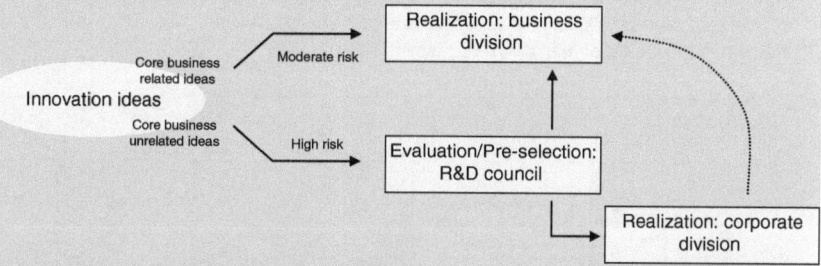

The R&D council is composed of R&D managers of the divisions, the representatives of the different technology and innovation management related corporate units, and leading representatives of the regional R&D centers. Its task with high risk innovation opportunities is twofold: on the one hand it preselects the interesting projects to be pursued and on the other hand it suggests if the project realization should take place on a divisional or corporate level, possibly with the involvement of further divisions.

Upon the suggestions of the R&D Council, the final decision of major projects to realize is up to the Board of Management.

Figure 9.17 Clariant: idea management and project realization adapted to different levels of risk and newness

**Topsoe: R&D activities according to different levels of market
and technology newness**

At Topsoe the distribution of R&D activities is fixed according to the level of technology and market newness*:

40% of R&D projects are run in existing markets on the one hand by fostering existing technologies and on the other hand by researching in emerging and new technologies. Research in emerging and new technologies is very goal oriented with specific deliverables and performances to achieve. These are set based on known weakness and problems of the existing technologies in their existing markets.

7% of the projects are conducted for existing markets. There are no reviewed deliverables or performance measurements. Thus research in this cluster is less problem solving focused.

30% of the projects are done in the field of so-called focused new ventures. The business opportunities behind these projects have been identified as highly innovative and market relevant. For most of the technologies developed in this cluster the proof of the concept is already done. They are focused toward deployment of a new market application.

10% of the projects are conducted in order to build up competencies in novel markets with novel technologies in exploratory, non-focused research.

The remaining 13% of the projects support the previous ones. They are conducted in the fields of licensing, patenting, etc.

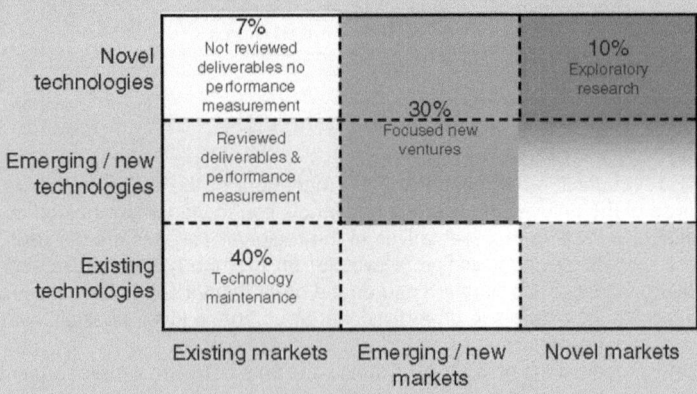

* Percentages are approximate figures, they are meant to provide an impression of the project's distribution and are subject to variance

Figure 9.18 Topsoe: R&D activities according to different levels of market and technology newness

HILTI: Differentiating four levels of technological Risks

In Hilti's innovation process the level of technological risk is one major criteria whether to realize/continue or to reject/stop an innovation project. Furthermore project goals and project realization forms in the innovation process differ depending on the level of risk.

When assessing risk in innovation projects Hilti applies four different risk measures with corresponding probabilities of success: Very high risk, high risk, moderate risk, and low risk.

Very high risk level	There are fundamental uncertainties in the physical processes on which the technology is based. It is for the moment unknown, whether a solution can be found. Probability of success: ca. 50%
High risk level	The principal physical aspects are known, but it is not fully clear, whether the mechanisms can be applied. It is open, whether a solution, transferable into a commercial product can be found. Probability of success: ca. 65%
Moderate risk level	The technological feasibility is verified in principle, but a limited number of problems are open and it is not fully clear, whether a solution for all of them can be found. Probability of success: ca. 80%
Low risk level	The concept of the technology is clear. The functional prototype is available. There are some technology engineering / design questions to be solved. It is, however, quite certain, that a solution can be found. Probability of success: ca. 90%

Figure 9.19   HILTI: differentiating four levels of technological risk

Philips: differentiating organizational realization forms of R&D projects

Philips differentiates the organizational realization of R&D projects according to a market criterion and a research type criterion. The market criterion distinguishes between existing markets and new markets. The research types are differentiated according the their application focus in products. There are three types: an early stage applied research, advanced stage applied research and product oriented research, and development. The matrix below visualizes the choice of the organizational realization depending on the combination of both criteria.

	Existing markets	New markets
Early stage applied research	Corporate research project	Corporate research project
Advanced stage applied research	Joint Project: corporate research & product division	Create a start-up company
Product oriented research and development	Product division project	--

Figure 9.20 Philips: differentiating organizational realization forms of R&D projects

Bayer: different R&D structures adapted to different types of innovation

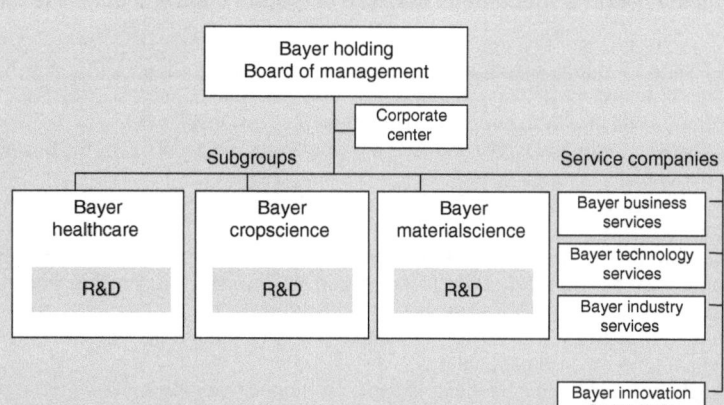

Since the latest corporate reorganization in 2002, Bayer has been structured in a hold-ing organization hosting three market oriented subgroups: Bayer healthcare AG, Bayer cropscience AG, and Bayer materialscience AG. All three subgroups are fully owned by the holding organization. Beside the subgroups, there are three service companies, also entirely owned by Bayer. They provide different kinds of functional services to the subgroups and to strategically relevant external customers. Bayer business services, Bayer technology services, and Bayer industry services are legally incorporated with their headquarters in Leverkusen. Furthermore, Bayer innovation GmbH is a subsidiary with headquarters in Düsseldorf, Germany. Its aim is to market concepts and innovative products to supplement Bayer's business portfolio, and to access new markets with strong growth potential, such as medical and security technology.

Directly subordinate to Bayer's Board of management, the so-called Corporate center comprises a number of staff functions. Among the staff functions within the Corporate center, Corporate development deals with the management of the current business portfolio elements and the planing of strategic innovation.

As R&D is performed decentralized, all three subgroups of Bayer, healthcare, crop-science and materialscience follow their own R&D activities. These are in the first place focused on applied research and (product)development fostering the existing core busi-nesses, but have also new business areas like "New Business Ventures" in Bioscience or the "New Business" center for materialscience. R&D efforts regarding the devel-opment and implementation of technology platforms spanning several subgroups that support the advancement of the company's core technologies and processes are taken care of by Bayer technology services.

Bayer innovation GmbH develops business ideas that cannot be related to existing businesses. The underlying technology of those businesses should however correlate with Bayer's existing competencies. Bayer innovation GmbH's mission is to develop new business opportunities by evaluation of possible new portfolio elements to fit with the corporate vision. The declared objective is to enrich and complete the holding's busi-ness portfolio and to penetrate into potential growth markets.

Figure 9.21 Bayer: different R&D structures adapted to different types of innovation

Syngenta: flexible cooperation between corporate units and business units

At Syngenta, many management activities, and among others, R&D activities, are run in interdisciplinary projects teams that are composed of people from the corporate level and from people from the business unit level.

The project team is typically created on the business unit level run by the business unit people.

These people form the core project team, they have the responsibility of the project.

The core project team has the possibility to access specific expertise from the corporate level. This expertise is organized in knowledge silos that comprise six global function pooling experts of specific knowledge backgrounds, such as research, regulatory, marketing, intellectual property, traits & genetics, and health assessment & environment safety.

People from these global functions can be assigned to temporarily support a core project team according to the required expertise.

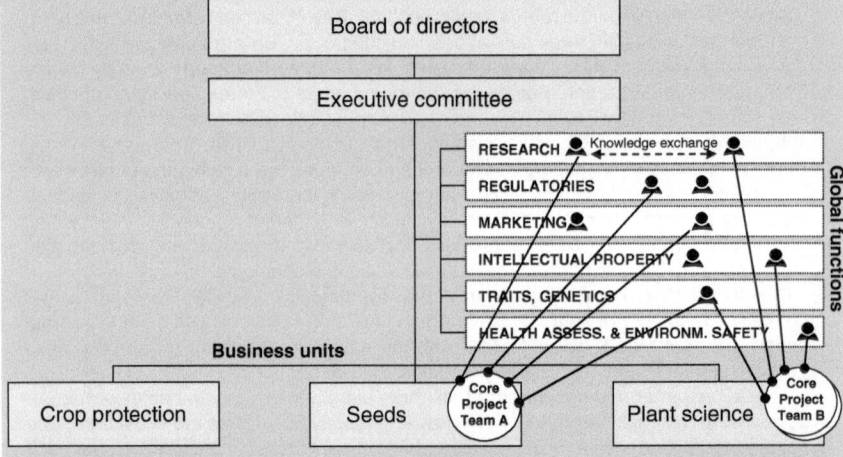

Pooling the experts at the corporate level in global functions has two significant advantages: first, various degrees of intensity in cooperation between the business unit level and the corporate level is facilitated as corporate experts can be deployed flexibly across the whole organization in one or more projects allowing a rapid compilation of project structures. Second, there is a great deal of knowledge exchange on the corporate level that the whole company can benefit from.

Figure 9.22 Syngenta: flexible cooperation between corporate units and business units

Degussa: collaboration of business and corporate unit in project houses

At Degussa the collaboration between business units and the corporate unit for high risk innovation projects exploring new technologies is processed through an organizational structure called the "project house". Project houses are a joint effort of business units together with Creavis, the operational arm of the corporate unit responsible for high risk innovation projects. It is a physical group of people composed of 20–30 company internal specialists collaborating with academia and external research institutions. Team members are delegated from their business units into a central location away from their organizational home unit. The participation as a team member is promoted by incentive plans.

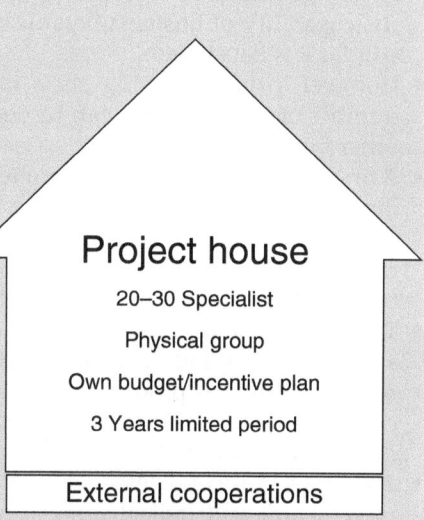

The project is managed and coordinated by Creavis. It has its own budget funded 50% by the participating business units and 50% by the corporate unit.

The primary goal of a project house is to develop new knowledge as a basis for new competencies in emerging technology areas that are of long term benefit for all participating business units. The newly developed knowledge as well as the infrastructure used in the project house is accessible company wide. One condition to start a project house is the participation and commitment of a significant number of Degussa's business units. This condition not only splits the financial commitment of the participating units, it also secures the market relevance of the endeavor as the interest of business units is a clear indicator of market relevance.

Project houses are run no longer than three years. After the end of this period, the team members including the corresponding infrastructure are reintegrated into their operational home ensuring the transfer of the newly elaborated knowledge. For newly developed knowledge with high market potential that does not fit into the existing businesses of the parent organization, there is the possibility for a new business.

Figure 9.23 Degussa: collaboration of business and corporate units in project houses

Managerial implications

- Coping with the indispensible business need to permanently innovate products, services, technologies, and entire business models is a highly demanding managerial task per se. Whereas the awareness and competence of companies to cope with incremental innovations is increasing, the capability of business organizations to handle radical innovations is still far less developed.
- However, from observing "best in class" companies in this respect, a number of lessons learnt can be derived which allow to be transferred to other business situations.
- A first lesson is telling, that – despite popular stories such as the unplanned discovery of the "post-it" product by 3M – the successful application of radical innovations does not result from coincidental circumstances, but rather from well reflected activities. And, such activities cannot be combined with ordinary product development processes.
- The reason for this is, given by the fact, that for product development and developing radical innovations two different time perspectives are relevant. Whereas product developments are ruled by a rather short time constant, discovering radical innovations consists of activities which are directed by a longer term time constant.
- Therefore, a second lesson learnt telling, that for the development of incremental and radical innovations two different – however related – processes have to be established.
- Both processes as such rely on the indispensible input from business intelligence systems. The process for radical innovations however contains at least two specifically new elements.
- The first element – and this is the third lesson learnt – can consist of being concerned with new competences. Example: the strategic plans of the Degussa company – being a chemicals company and therefore a "commodity producer" – contains explicitly the issue "competence planning". This issue is dealing with future innovation fields of strategic significance and correspondingly new areas of knowledge, such as nanotechnology, catalysis, polymers, fermentation, etc. Having decided on the priorities of these areas, so-called "Project Houses" are established. They represent projects which assemble internal and external professionals who deepen their knowledge and expertise in the new field in a location separate from their usual work places. Such projects are limited to exactly three years and are financed by equal resources from corporate and divisional origin. After the three years, the internal professionals return to their original work places and now try to apply the newly gained knowledge for fundamentally new applications.
- The second element – to be considered as fourth lesson learnt – consists of including venture capital activities. This can mean a variety of options,

such as participation or even ownership of venture capital companies, establishing internal start-up companies, and in particular running "start-up watch lists". This element of modern innovation management is aiming at providing transparency of the relevant external "start-up scene" in order to be in the position to conduct so-called "trade deals" which means buying suitable start-up companies.

- This engagement in the "start-up scene" is indeed a highly recommendable strategic option for established companies: From the point of view of such companies, start-ups do represent first "outsourced" technology risks, second, "outsourced" marketing risks, and third gained developing time – all three factors being significant business advantages. They all are in direct contrast to efforts which are directed towards operating internal R&D structures focused on developing radically new technologies and innovations. In other words: a proactive and systematic involvement in the "start-up scene" provides most valuable strategic flexibility.

Notes

1. A technological paradigm is a model and solution approach of specific technological problems. It is based on determined principles of scientific disciplines and specific material technologies (Dosi, 1982: 152).
2. Recently authors have suggested differentiating further the competence destroying and competence enhancing concept stating that they are composed of two distinct constructs that, although correlating, separately characterize an innovation: new competence acquisition and competence enhancing / destroying (Gatignon et al., 2002: 1103).
3. Projects with a 5–10 times improvement in performance, 30–50 percent reduction in cost and/or new-to-the-world performance (Rice et al., 1998: 52).
4. A product's technology base includes all the technologies – specific areas of technical competence – which are needed to design and produce the product (Ehrnberg and Jacobsson, 1993: 28).
5. Exception to this market perspective is the description from Green et al. (1995: 203ff).
6. Not all the findings point in the same direction, for example: Herstatt and Lettl (2004) describe a case where customers were successfully involved in the development of radical innovation products, see also von Hippel (1988).
7. Different authors developed technology life cycle models: Little, A. (1998), Roussel, P. A. (1984), Ford and Ryan (1981).
8. For more details on the generic strategy development process please refer for example to (Tschirky, 2003: 59).
9. Markham and Griffin (1998: 436ff) have evidence that questions the generally positive influence of champions for innovation projects. For further reading about champions refer also to Howell and Higgins (1994), Maidique (1980) and Markham (1998).
10. The 'not invented here syndrome' is an attitude of employees that prevent them from accepting ideas and innovations that have been developed outside of their own company (Servatius, 1988: 14).

11. For a detailed description of corporate acquisitions in innovation driven industries see Bannert (2004).

References

Abernathy, W. J. and Clark, K. B. (1985). "Innovation: Mapping the Winds of Creative Destruction." *Research Policy* 14: 3–22.

Abernathy, W. J. and Clark, K. B. (1991). "Innovation: Mapping the Winds of Creative Destruction." In M. L. Tushman and W. L. Moore (Eds), *Readings in the Management of Innovation* (2nd edition). New York: 55–78.

Abernathy, W. J. and Utterback, J. M. (1978). "Patterns of Industrial Innovation." *Technology Review* 6/7: 41–7.

Afuah, A. (1998). *Innovation Management: Strategies, Implementation, and Profits.* Oxford: Oxford University Press.

Anderson, P. and Tushman, M. L. (1990). "Technological Discontinuities and Dominant Design: A Cyclical Model of Technological Change." *Administrative Science Quarterly* 35: 604–33.

Bannert, V. (2004). *Managing Corporate Acquisitions in Innovation Driven Industries – Bringing Technology into Due Diligence.* Zürich: Verlag Industrielle Organisation.

Bower, J. L. and Christensen, C. R. (1995). "Disruptive Technologies: Catching the Wave." *Harvard Business Review* (January-February): 43–53.

Bower, J. L. and Christensen, C. R. (1995a). "Technisch revolutionäre Produkte: Wenn die Stammkunden mauern." *Harvard Business Manager* 3: 88–96.

Brodbeck, H., Bucher, P., Birkenmeier, B., and Escher, J.-P. (2003). "Evaluating and Introducing Disruptive Technologies." In H. Tschirky (Ed.), *Technology and Innovation Management on the Move.* Zürich: Verlag Industrielle Organisation: 137–51.

Christensen, C. M. (1997). *The Innovators Dilemma, When New Technologies Cause Great Firms to Fail.* Boston, MA: Harvard Business School Press.

Christensen, C. M., Johnson, M. W., and Rigby, D. K. (2002). "Foundations for Growth: How to Identify and Build Disruptive New Businesses." *MIT Sloan Management Review* 43: 22–31.

Christensen, C. M. and Raynor, M. E. (2003). *The Innovator's Solution.* Boston, MA: Harvard Business School.

Clark, G. (2003). "The Disruption Opportunity." *MIT Sloan Management Review* (summer): 27–32.

Clark, G. and Bower, J. L. (2002). "Disruptive Change." *Harvard Business Review* May: 3–8.

Damanpour, F. (1988). "Innovation Type, Radicalness, and the Adoption Process." *Communication Research* 15(5): 545–67.

Dosi, G. (1982). "Technological Paradigms and Technological Trajectories." *Research Policy* 11: 147–62.

Ehrnberg, E. and Jacobsson, S. (1993). "Technological Discontinuity and Competitive Strategy – Revival through FMS for the European Machine Tool Industry?." *Technological Forecasting and Social Change* 44: 27–48.

Ehrnberg, E. and Sjöberg, N. (1995). "Technological Discontinuities, Competition and Firm Performance." *Technology Analysis and Strategic Management* 7(1): 93–105.

Floyd, C. (1996). "Managing Technology Discontinuities for Competitive Advantage." *Prism* (2nd Quarter): 5–21.

Ford, D., and Ryan, C. (1981). "Taking technology to Market." *Harvard Business Review* 59(2): 117–26.

Foster, R. N. (1986). Die *technologische Offensive.* Wiesbaden: Gabler.

Foster, R. N. and Kaplan, S. (2002). *Schöpfen und Zerstören*. Wien / Frankfurt: Redline Wirtschaft.

Garcia, R. and Roger, J. C. (2002). "a Critical Look at Technological Innovation Typology and Innovativeness Terminology: a Literature Review." *Journal of Product Innovation Management* 19: 110–32.

Gatignon, H., Tushman, M. L., Smith, W., and Anderson, P. (2002). "A Structural Approach to Assessing Innovation: Construct Development of Innovation Locus, Type and Characteristics." *Management Science* 48(9): 1103–22.

Gilbert, C. (2003). "The Disruption Opportunity." *MIT Sloan Management Review* (Summer): 27–32.

Green, S. G., Gavin, M. B., and Aiman-Smith, L. (1995). "Assessing a Multidimensional Measure of Radical Technological Innovation." *IEEE Transactions on Engineering Management* 42(3): 203–14.

Herstatt, C. and Lettl, C. (2004). "Management of 'technology push' development projects." *International Journal of Technology Management* 27(2/3): 155–75.

Howell, J. M. and Higgins, C. A. (1994). *Champions for the Future*. Boston, MA: Harvard Business School Press.

Hruby, J., Kassicieh, S. K., and Walsh, S. T. (2000). "Commercialization of Disruptive Technologies: The Process of Discontinuous Innovations." *IEEE 2000*: 335–9.

Jolly, V. K. (1997). *Commercializing New Technologies*. Boston, MA: Harvard Business School Press.

Kassicieh, S. K., Walsh, S. T., Cummings, J. C., McWhorter, P. J., Romig, A. D., and Williams, W. D. (2002). "Factors Differentiating the Commercialization of Disruptive and Sustaining Technologies." *IEEE Transactions on Engineering Management* 49 (4, November): 375–87.

Kessler, E. H. and Chakrabarti, A. K. (1999). "Speeding Up the Pace of New Product Development." *Journal of Product Innovation Management* 16: 231–47.

Kim, W. C. and Mauborgne, R. (1997). "Value Innovation – The Strategic Logic of High Growth." *Harvard Business Review* (January-February): 103–12.

Kostoff, R. N., Boylan, R., and Simons, G. R. (2004). "Disruptive Technology Roadmaps." *Technological Forecasting and Social Change* 71: 141–59.

Kusunoki, K. (1997). "Incapability of Technological Capability: A Case Study on Product Innovation in the Japanese Facsimile Machine Industry." *Journal of Product Innovation Management* 14: 368–82.

Lambe, C. J. and Spekman, R. E. (1997). "Alliances, External Technology Acquisition, and Discontinuous Technological Change." *Journal of Production and Innovation Management* 14: 102–16.

Learned, E. P., Christensen, C. R., Andrews, K. R., and Guth, W. D. (1965). *Business Policy: Text and Cases*. Irwin, IL: Homewood.

Lehmann, A. (1994). *Wissensbasierte Analyse technologischer Diskontinuitäten*. Wiesbaden: Deutscher Universitätsverlag.

Leifer, R. (1998). "An Information Processing Approach for Facilitating the Fuzzy Front End of Breakthrough Innovations." *Conference Proceedings IEMC 1998*: 130–5.

Leifer, R. (2000). *Radical Innovation*. Boston, MA: Harvard Business School Press.

Leifer, R., McDermott, C. M., O'Connor, G. C., Peters, L. S., Rice, M. P., and Veryzer, R. W. (2000). *Radical Innovation*. Boston, MA: Harvard Business Review Press.

Linton, J. D. (2002). "Forecasting the Market Diffusion of Disruptive and Discontinuous Innovation." *IEEE Transactions on Engineering Management* 49(4): 365–74.

Little, A. (1998). *Innovation als Führungsausgabe*. Frankfurt: Campus Verlag.

Lynn, G. S., Morone, J. G., and Paulson, A. S. (1996). "Marketing and Discontinuous Innovation." *California Management Review* 38(3): 8–37.

Maidique, M. A. (1980). "Entrepreneurs, Champions, and Technological Innovation." *Sloan Management Review* 21(2): 59–76.

Maidique, M. and Hayes, R. (1984). "The Art of High Technology Management." *Sloan Management Review* 25(2): 17–32.

Maidique, M. A. and Patch, P. (1982). *Corporate Strategy and Technological Policy*. Cambridge, MA: Ballinger.

Markham, S. A. (1998). "Longitudinal Examination of How Champions Influence Others to Support Their Projects." *Journal of Product Innovation Management* 15: 490–504.

Markham, S. K. and Griffin, A. (1998). "The Breakfast of Champions: Associations between Champions and Product Development Environments, Practices and Performance." *Journal of Product Innovation Management* 15: 436–54.

Mascitelli, R. (2000). "From Experience: Harnessing Tacit Knowledge to Achieve Breakthrough Innovation." *Journal of Product Innovation Management* 17: 179–93.

McDermott, C. M. and O'Connor, G. C. (2002). "Managing Radical Innovation: An Overview of Emergent Strategy Issues." *Product Innovation Management* 19: 424–38.

McKelvey, M. D. (1996). "Discontinuities in Generic Engineering for Pharmaceuticals? Firm Jumps and Lock-in in Systems of Innovation." *Technology Analysis and Strategic Management* 8(2): 107–16.

Milliken, F. J. (1990). "Perceiving and Interpreting Environmental Change: An Examination of College Administrators' Interpretation of Changing Demographics." *The Academy of Management Journal* 33(1): 42–63.

Mintzberg, H. and Waters, J. A. (1985). "Of Strategies, Deliberate and Emergent." *Strategic Management Journal* 6: 257–72.

Mokyr, J. (1990). *The Level of Riches*. New York: Oxford University Press.

Nayak, P. R. and Ketteringham, J. M. (1986). *Breakthroughs!*. New York.

North, K. and Tucker, S. (1987). *Implementing Routine and Radical Innovations*. Lexington: Free Press.

O'Connor, G. C. (1998). "Market Learning and Radical Innovation: A Cross Case Comparison of Eight Radical Innovation Projects." *Journal of Product Innovation Management* 15: 151–66.

O'Connor, G. C. and Veryzer, R. W. (2001). "The Nature of Market Visioning for Technology-Based Radical Innovation." *The Journal of Product Innovation Management* 18: 231–46.

O'Reilly, C. and Tushman, M. L. (2004). "The Ambidextrous Organization." *Harvard Business Review* (April): 74–81.

Osterloh, M. and Wartburg, I. von (1998). "Organisationales Lernen und Technologie-Management." In H. Tschirky and S. Koruna (Eds), *Technologie-Management, Idee und Praxis*. Zürich: Orell-Füssli: 138–56.

Peter, L. and O'Connor, G. C. (2002). "Building the Capacity for Change: Radical Innovation and New Product Development in the Multinational Firm." *IEEE 2002*: 612–16.

Peters, L. (2000). "The Role of the Research University in MNC Radical Innovation (RI)." *IEEE 2000*: 669–73.

Peters, L. S. (1996). "The Virtual Enterprise and the Source of Technology in Discontinuous Innovation." *IEMC 1996*: 470–4.

Porter, M. E. (1998). "Competitive Strategy in emerging Industries." In H. Mintzberg and J. B. Quinn (Eds), *Readings in the Strategy Process: Concepts and Contexts*. New Jersey: Prentice-Hall.

Prahalad, C. K. and Bettis, R. A. (1986). "The Dominant Logic: A New Linkage between Diversity and Performance." *Strategic Management Journal* 7: 485–501.

Rafii, F. and Kampas, P. J. (2002). "How to Identify Your Enemies before They Destroy You." *Harvard Business Review* (November): 115–23.

Rice, M. P. (1996). "Virtuality and Uncertainty in the Domain of Discontinuous Innovation." *IEMC* 1996: 528–32.

Rice, M. P., Kelley, D., Peter, L., and O'Connor, G. C. (2001). "Radical Innovation: Triggering Initiation of Opportunity Recognition and Evaluation." *R&D Management* 31(4): 409–20.

Rice, M. P., Leifer, R., and O'Connor, G. C. (1996). "Managing the Transition of Discontinuous Innovation Project to Operational Status." IEEE Conference Proceedings: 586–90.

Rice, M. P., O'Connor, G. C., Peter, L., and Morone, J. G. (1998). "Managing Discontinuous Innovation." *Research Technology Management* (May/June): 52–8.

Roussel, P. A. (1984). "Technological Maturity Proves a Valid and Important Concept." *Research Management* 27: 29–34.

Salavou, H. and Lioukas, S. (2004). "Radical Product Innovations in SMEs: The Dominance of Entrepreneurial Orientation." *Creativity and Innovation Management* 12(2): 94–106.

Savioz, P., Lichtenthaler, E., Birkenmeier, B., and Brodbeck, H. (2002). "Organisation der frühen Phasen des radikalen Innovationsprozess." *Die Unternehmung* (06/2002): 393–407.

Servatius, H.-G. (1988). *New Venture Management*. Wiesbaden: Betriebswirtschaftlicher Verlag Gabler GmbH.

Shenhar, A. J., Dvir, D., and Shulman, Y. (1995). "a Two Dimensional Taxonomy of Products and Innovations." *Journal of Engineering and Technology Management* 12: 175–200.

Song, X. M. and Montoya-Weiss, M. M. (1998). "Critical Development Activities for Really New versus Incremental Products." *Journal of Product Innovation Management* 15: 124–35.

Stoelhorst, J. W. (2002). "Transition Strategies for Managing Technological Discontinuities: Lessons from the History of the Semiconductor Industry." *International Journal of Technology Management* 23(4): 261–86.

Strebel, P. (1992). *Breakpoints, How Managers Exploit Radical Business Change*. Boston, MA: Harvard Business School Press.

Strebel, P. (1995). "Creating Industry Breakpoints: Changing the Rules of the Game." *Long Range Planning* 28(2): 11–20.

Suarez, F. F. and Utterback, J. M. (1995). "Dominant Design and the Survival of Firms." *Strategic Management Journal* 16: 415–30.

Tidd, J. (1995). "Development of Novel Products through Intraorganizational and Interorganizational Networks." *Journal of Product Innovation Management* 12: 307–22.

Tschirky, H. (2003). "The Technology Awareness Gap in General Management." In H. Tschirky, H.-H. Jung, and P. Savioz (Eds), *Technology and Innovation Management of the Move*. Zürich: Verlag Industrielle Organisation: 21–41.

Tushman, M. L. and Anderson, M. (1986). "Technological Discontinuities and Organizational Environments." *Administrative Science Quarterly* 31(3): 439–65.

Tushman, M. L., Anderson, P. C., and O'Reilly, C. (1997). "Technology Cycles, Innovation Streams, and ambidextrous Organizations: Organization Renewal through Innovation Streams and Strategic Change." In M. L. Tushman and P. I. Anderson (Eds), *Managing Strategic Innovation and Change*. New York: Oxford University Press: 3–23.

Tushman, M. L. and Nadler, D. (1986). "Organizing for Innovation." *California Management Review* 28: 74–92.

Tushman, M. L. and O'Reilly, C. A. (1996). "Ambidextrous Organizations: Managing Evolutionary and Revolutionary Change." *California Management Review* 38(4): 8–30.

Tushman, M. L. and O'Reilly, C. A. (1996a). "Ambidextrous Organizations: Managing Evolutionary and Revolutionary Change." In R. A. Burgelman, M. A. Maidique, and S. C. Wheelwright (Eds), *Strategic Management of Technology and Innovation*, (3rd edition). Boston, MA: McGraw-Hill: 724–37.

Tushman, M. L. and O'Reilly, C. A. (1998). "Unternehmen müssen auch den sprunghaften Wandel meistern." *Harvard Business Manager* 1: 30–44.

Utterback, J. M. (1994). *Mastering the Dynamics of Innovation, How Companies Can Seize Opportunities in the Face of Technological Change*. Boston, MA: Harvard Business School.

Veryzer, R. W. (1998a). "Discontinuous Innovation and the New Product Development Process." *Journal of Product Innovation Management* (15): 304–21.

Veryzer, R. W. J. (1998). "Key Factors Affecting Customer Evaluation of Discontinuous New Products." *Journal of Product Innovation Management* 15: 136–50.

Vojak, B., A. and Chambers, F. A. (2004). "Roadmapping Disruptive Technical Threats and Opportunities in Complex, Technology-Based Subsystems: the sails Methodology." *Technological Forecasting and Social Change* 71: 121–39.

von Hippel, E. (1988). *The Sources of Innovation*. New York: Oxford University Press.

Walsh, S., Kirchhoff, B. and Newbert, S. (2002). "Differentiating Market Strategies for Disruptive Technologies." *IEEE Transactions on Engineering Management* 49(4): 341–51.

Wieandt, A. (1995). "Zur Entstehung von Märkten durch Innovationen." *Betriebswirtschaftliche Forschung und Praxis* 47(4): 447–71.

Wood, S. C. and Brown, G. S. (1998). "Commercializing Nascent Technology: The Case of Laser Diodes at Sony." *Journal of Product Innovation Management* 15: 167–83.

10
Market Research for Radical Innovations – Lessons from a Lead User Project in the Field of Medical Products

Cornelius Herstatt

Introduction

In order to ensure their long-term growth, companies require a balanced portfolio of both incremental and more significant innovations. In the ideal situation, the continuous improvement of the existing product/service range would provide the financial support for the ongoing development of the more significant innovations.

In successful companies, the development of incremental innovation projects belongs to the day-to-day routine work. Management possesses an arsenal of methodologies, with which these projects can be systematically planned, steered, and controlled. Traditional market research therefore offers many possibilities to determine the needs of the customer and to test the developed concepts in the target markets prior to market launch.

However, it is a very different situation for so-called "breakthrough" innovation projects. In such cases, the results from market research studies continually evoke disappointment, since it mostly appears impossible to determine the demands of tomorrow's markets applying classical need-assessment methods. This may be due to certain methodologies providing customers with only limited opportunity to articulate innovative ideas. It can however also be due to the typical customer involved in market research studies. It is rare that sample testing of the current market leads to the identification of potential starting points for innovations due to the average customer not being capable of looking beyond what is currently on offer in the marketplace.

Leading companies such as 3M, HILTI, Nortel Networks or Kelloggs are increasingly working with Lead Users in the early phases of innovation projects (Herstatt and Lüthje, 2003). Lead Users are especially well-qualified, advanced users who are both sufficiently well motivated and qualified to make significant contributions to the development of new products or services (von Hippel, 1988). Their identification and involvement is supported through a specific process. This approach was originally sketched by MIT Professor Eric von Hippel, has been used successfully in various innovation projects and was further developed by von Hippel and his scholars since then (www.leaduser.com).

The Lead User approach and its implementation in commercial innovation practice forms the basis of this paper. After a short description of Lead Users, we will show that innovative customers exist in numerous product fields. Following this, the process steps of the Lead User Approach will be outlined briefly. Additionally, the functionality and effectiveness of the approach will be shown and discussed. Further we describe the findings of a case study in which innovation concepts for surgical hygiene and sterile products are developed through the involvement of Lead Users. Finally, we discuss the experience collected with the Lead User approach.

What are Lead Users?

Even an intensive integration of customers for the purpose of innovation often results in disappointment, as the outcomes from such studies are reported to be only of marginal value. But why? These studies often primarily result in individual ideas for smaller, incremental improvements of the existing product program and only rarely in applications for significant new innovations. Companies that depend exclusively on the results of such studies run the risk that their products will become outdated and over the long-term lose market share.

Most of the typically "representative" customers involved in such need-assessment surveys just appear to be unable to free themselves from the current situation and to imagine the needs of the future that foreshadow market development. Their experience with the current market products on offer prevents them from thinking creatively about future possibilities ("functional fixedness") (von Hippel, 1988).

According to von Hippel, it appears to make much more sense to seek the cooperation of Lead Users (von Hippel, 1988) for such projects. Lead Users are identifiable predominately through two characteristics:

- Lead Users anticipate the future needs of the market and do so significantly earlier than the majority of other customers.
- Lead Users profit strongly from innovations that solve their problems or enable new opportunities.

The first characteristic is the basis for the qualification of the Lead User to possess the ability to make valuable contributions to the development of new products. Through their prominent role in the recognition of new challenges and application problems, manufacturers can use them as predictors of the needs of the "market-of-tomorrow." As opposed to representative customers, Lead Users are much better in the position to imagine themselves in the future application compared to ordinary users. They are able to achieve this from their day-to-day experience and their strong dissatisfaction with current market offers.

The second Lead User characteristic is about the motivation for innovation. The expected benefits associated with an innovation can become so strong that the Lead User become innovators themselves, since they can not be satisfied by manufacturers and the existing product offers on the market. This "forced" entrance into innovation can occur quite often due to manufacturers either not being aware of, or not appreciating, the importance of the emerging needs of smaller customer groups.

Examples of Lead User innovations

The existence of Lead Users can be observed in numerous examples of innovations that were initiated by users of products and services, often without any involvement of a manufacturer. Such User innovations occur in both consumer and in industrial goods markets.

Consumer goods

A well-known example of an innovation that stemmed from a consumer is "TipEx," invented at the end of the 1950s by a secretary. The invention was later taken over by 3M and implemented on an industrial scale. The sport's drink "Gatorade" was developed by the trainer of a college football team. In general, the recreation and sport markets are rich in User innovations. Newly developed sports usually stem from the participants themselves. In an investigation of innovations in skate boarding, snow boarding, and surfing it was shown that practically all of the basic product development was carried out by the participants and not sports article manufacturers (Shah, 2000; Tietz, 2003). In a study in the area of outdoor and trekking products, a high level of innovative product users was also confirmed (Lüthje, 2000).

Industrial goods

In some industrial goods markets users are also responsible for important innovations. For example, it was shown with semiconductor and capacitor manufacturing that significant advances in technology were made by the semiconductor manufacturers themselves and not from the developers of the respective process technologies (von Hippel, 1977). Similar results were found for other process technologies such as CAD and CAM systems. Innovative product users were likewise found in the medical field. Clinics and doctors in many different fields are responsible for many new developments (Shaw, 1985), and the market for medical products is rich of user innovations (Lüthje, 2003b). Further, we could prove the existence of Lead Users for the fastening industry in construction (Herstatt, 1994).

Even in the dynamic IT industry there are numerous examples of innovations to be found that have been developed by Lead Users. The operating system "Linux" or the server software "Apache" are two prominent examples of the "Open Sources Movement." These projects were initiated by individual software users – Linus Tovalds in the case of Linux or Rob McCool at Apache – and accepted later from important user groups. These "User Communities" improve and test the programs and decide independently over the recording of new program codes in the software. Even today when the marketing of the products is partly taken over by companies (for example Red Hat Inc., VA Linux Systems), the manufacturers are rarely involved in the development of the software.

Existing research shows that innovation activities are not limited to a small section of users in a market. It can be shown that the percentage of

Table 10.1   Fraction of users who build solution for own use within different user populations

Study	Field of innovation	Users sampled (n)	% of users who developed solution for own use
Lüthje (2003b)	Medical surgery equipment (Germany)	261	22%
Lüthje (2003a)	Equipment for outdoor sports (Germany)	153	10%
Franke and Shah (2002	"Extreme" sporting equipment (Germany)	197	38%
Tietz (2003)	Kite surfing equipment (Australia)	157	26%
Lüthje, Herstatt, and von Hippel (2002)	Mountainbike equipment (USA)	287	19%
Morrison, Roberts, and von Hippel (2000)	Library information search system OPAC (Australia)	102	18%
Herstatt and von Hippel (1992)	Pipe hangers hardware (Switzerland)	74	36%
Urban and von Hippel (1988)	PC-CAD for the design of printed circuit boards (USA)	136	24%

users in different industries who improve prototypes or develop completely new solutions is considerable. The numbers vary between 10 percent and nearly 40 percent. When manufacturers make concerted efforts to search for users who have already developed solutions for their needs, a rich source of innovative ideas can be the reward. Table 10.1 shows some of the recent studies carried out in the field of user innovation.

The examples of User innovations in different industries and product fields raise the question of how a manufacturer can best profit from these activities for the purpose of innovation. If Lead Users are identified and included in innovation projects early, manufacturers have a good chance to profit from the immense potential of such users.

Targeting Lead Users and their involvement in innovation projects

In the 1980s von Hippel developed a systematic approach to search for Lead Users. The so-called Lead User process has since, on the basis of numerous applications, been detailed and developed further (Herstatt, 2002). The methodology is based on a four-phase process that begins with the delineation of a search field and ends with the development of a product concept. Applications of the methodology by leading manufacturers until now have lasted for periods of between four and nine months.

The Lead User process

Phase I: Start of the Lead User project

A Lead User project is too demanding for it to run within a company's functional area alongside routine duties. Hence, the creation of a dedicated, interdisciplinary team consisting of people from marketing, sales, R&D, and production is required. The team can encompass three to six members where a minimum of 50 percent of their work time should be available for the project.

The project team begins with delineation, as accurately as possible, of the search field that is a market, product field or service area in which innovative ideas could be found. Goal formulation then follows with the determination of the basic demands required to satisfy the development (desired degree of innovation). Both internal and external influences on the project should be considered at this stage.

Phase II: Trend prognosis

The identification of critical trends and developments in the search field is important in order to confirm that the Lead Users detect market needs earlier than other customers. These include technology and market trends relating to each search field as well as the predicting of economic, legal, and business developments. When such trends are known, Lead Users who lead these trends can be searched for. In this sense the identified trends turn out to be a key input-variable for a project, since these will guide the direction of all the activities that will follow.

There are many different sources of information available for trend analysis. In addition to the analysis of secondary source information (for example academic publications, data banks, internet), presentations by experts have proved to be especially valuable. The choice of experts should include a wide variety of expert knowledge in order that important developments are not missed (for example concurrent technologies, newly created markets). Often, organizing a few interviews with experts can allow a good overview of important developments of the investigated markets. At this point the first concrete indications of suitable Lead Users appear as often the interviewed experts are a good source for identifying advanced customers.

Phase III: Identification of Lead Users

In order to ascertain who the Lead Users are, the project team must first determine the indicators that will allow for their correct identification. Amongst other considerations, it is important that the Users actually do lead the trends that were chosen as being important in the previous phase. Further indicators that have proved worthwhile in previous applications of the methodology are; the dissatisfaction with the current market offerings, high understanding of problems in the search field in question, and the ability to develop their own product improvements or solutions.

The process of searching for the Lead Users itself is a creative one that must be tailored to the specific conditions of the relevant search field. It is possible to follow three basic process types for this search:

Screening Approach: For a large number of product users "screening" for the presence of the previously identified indicators can be carried out for Lead Users. As well as the information freely available about companies from customer data banks, customer complaints or externally produced reports, telephone interviews can also be used. This approach is practical when the number of customers in the market is overseeable and a complete screening of all users possible.

Networking Approach: This approach involves a small number of customers who are asked if they know of other product users, who have new needs or are innovatively active. These kinds of recommendations usually lead very quickly to interesting Lead Users. A significant advantage of this methodology lies in that the team is often exposed to analogous fields in which similar challenges are present as in the search field in question. An example for this is the Lead User project for "medical imaging" for the recognition of small tumors. The search process here didn't just involve leading radiologists but also experts from the military field as Lead Users. The identification of minute details by the military (for example weapons) on satellite images is achieved through pattern-recognition software that is capable of very good results when the resolution is particularly poor. The idea of using a pattern-recognition system for "medical imaging" was completely new as previously the focus had always been on improving resolution.

Phase IV: Development of product concepts

With the identification of a Lead User group, the project team has often already collected some ideas for innovations, albeit vague ones. The Lead Users are then brought together in a workshop that typically lasts two to three days in order to develop the ideas further and to combine them. The workshop serves as an efficient discussion platform where the product users as well as some of the company staff should also be involved.

Depending on the search field, the issue of Intellectual Property Rights will probably need to be addressed. It is sensible, as a minimum, for the parties involved to sign a confidentiality agreement and also to determine the rights of ownership and use. From experience, Lead Users are generally prepared to absolve the rights to ideas without any significant form of compensation.

The workshop revolves around the development of problems associated with existing market offerings and the challenges of future solutions. The results of these discussions, aided by creativity and workshop techniques, serve as the basis for the subsequent development of concrete innovation ideas in teams of three to five people. Throughout these processes, the ideas are fleshed out such that by the end they form a basic sketch, concept outline or model.

After completion of the workshop, the ideas are subjected to a preliminary evaluation via their presentation to the respective decision maker from within the organization. The team members act as product champions

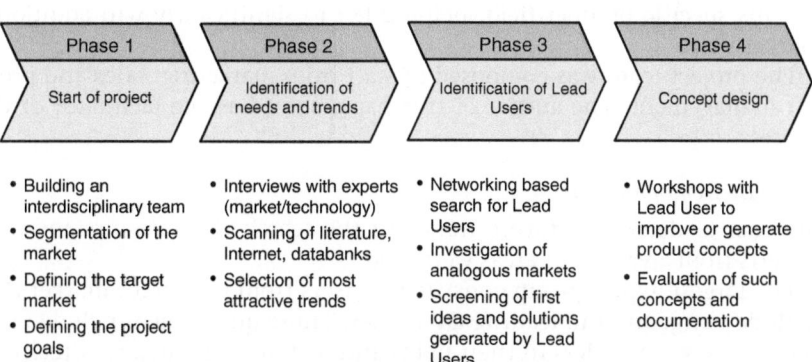

Figure 10.1　The Lead User process

during this process. When the decision is made for the further development of a concept, the normal development and evaluation processes that the organization uses to pursue innovation ideas are applied. The Lead User approach is therefore no substitute for traditional innovation management or market research techniques.

Figure 10.1 summarizes the various steps of the Lead User approach.

Lead Users create new product lines: the example of Johnson & Johnson Medical, Germany

A recent Lead User project was carried out by the German subsidiary of Johnson & Johnson Medical (referred to as J&J). The management staff of surgical hygiene products, including disposable articles used during surgery (gowns, masks, and garments), saw the Lead User approach as a promising approach for the development of ideas for completely new products. An additional motivator for this decision was that a successful application of the Lead User approach at 3M-Medical division was reported about in the literature (Thomke et al., 1998). J&J Management was curious to find out if the Lead User approach was helpful to identify concepts for breakthrough innovations.

Phase I: Start of the Lead User project

The key decision maker at J&J chose the patient coverings and the protective clothing of operation personnel as the search fields for which innovative ideas should be sought. This product area is in the very mature phase of the product life cycle. Significant differences between competing products are today not really present, so that the customer's choice is based basically on the price only. It is likely that the price competition increases further due to the increasing cost pressures being experienced by hospitals and clinics. Knowing this, there was only a little hope at the beginning of the project that the Lead User approach would allow the generation of really new ideas for this specific product field, helping J&J to significantly win additional market share.

The project team was composed of staff from marketing, sales and product management. The author of this paper served as the facilitator of the process.

Phase II: Trend prognosis

In order to determine future trends in the search field, discussions with experts from a variety of areas were held:

The largest expert group consisted of users from the target market that includes surgeons and leading OP nurses. Those questioned included surgeons who were leaders in their fields and had implemented new technology in their surgery ("leading edge users"). Also included were "extreme users." This group consisted of doctors who worked under particularly difficult conditions such as in the tropics (poor hygiene controls, poor product availability) or burns surgeons (long operations with large, moist wounds).

In addition to current product users, other experts from the target market were involved in the discussions (a second expert group). This included hygiene experts, buyers in clinics or logistics personnel from hospitals.

The third group was made up from experts in analogous application fields, primarily from semiconductor production. This is due to the conditions required in the "clean rooms" of chip factories being similar to operation rooms (free from dust or particles).

From these expert discussions alone a large number of ideas for the improvement of the coverings and protective clothing resulted. This outcome, due to the incremental nature of those ideas, whilst not the key target of this project, was however considered to be a useful byproduct by J&J. The discussions also resulted in a deeper understanding about developments that would play an important role in this search field for the future. The bandwidth ranged from economical trends (the need for the reduction in hospital inventory), to medicinal developments (the increase of particular infections) to concrete technological trends (new surgical techniques). In light of this great range, focusing on one idea was required.

After this, The project team decided to concentrate on one specific technological trend, "surgical robotics." In some surgical fields (like hip and knee prosthetics, minimally invasive heart surgery, brain surgery) surgical robots are already applied today. They support the surgeon in the tasks that require the highest precision possible such as with exact drilling of bones. Therefore associated with the increasing use of this technology are completely new demands for the protection against infection in OP that in turn allow the possibility for offering new hygiene products. It is here that a great chance for innovations was seen. For this project, hip and knee prosthetics were chosen from the aforementioned application field. Due to the robotic systems developed for this field, "CASPAR" and "ROBODOC," already having some distribution, any product improvement in these systems, at least in the middle term, promised a high market potential.

Although another J&J branch is active in the field of surgical robotics as they design components, the hygienic problems of these had not been considered as an attractive field for innovation by J&J. This was a positive surprise and created confidence that the Lead User approach was also valuable in the context of this project.

Phase III: Identification of Lead Users

Lead Users were sought after in surgeons that primarily operate in hip surgery. Within this field users were identified who worked at the

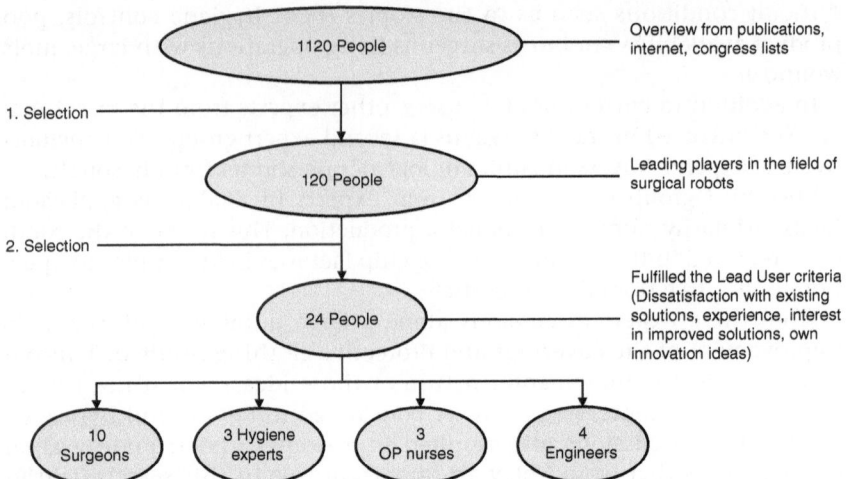

Figure 10.2   Selection process for identifying Lead Users

forefront of technology, identified serious hygienic problems, reported making use of improvements made in this field, and even possibly made their own attempts to solve these problems. At this point an exhaustive screening process was applied. Through research of relevant publications and in the internet, as well as attending conferences and interviews with experts, around 130 people were identified who play a critical role in robotic surgery today. These people were contacted via telephone or visited in their clinic and questioned. Furthermore, those selected were asked to name any other people that were leaders and active in an innovative way in robotics or similar fields. These people were also contacted ("networking approach"). The systematic selection process of Lead User is outlined in Figure 10.2.

During the selection process, a group of 20 people were identified who matched the Lead User criteria. The members of this group stemmed from four fields displaying varying competencies; the surgeons as users of robotic systems, the OP personnel as the people responsible for hygienic conditions in OP (for example covering of the robot and patient), the clinical engineers as being responsible for the technical maintenance and the clinical hygiene experts for infection-related questions.

Phase IV: Development of the product concept

From these 20 Lead Users, 18 were present for the subsequent two-day Lead User workshop. This number of participants reflected the high degree of involvement of the group in the development of the product concept. The group of Lead Users was complemented by three robot-manufacturing representatives, three employees from J&J, and through the author of this paper. An initial priority of the workshop was the agreement of the participants to surrender any user rights to J&J.

The group started the workshop with the aim of developing concrete concepts for innovative hygiene products in the field of robot-assisted surgery. Initially the participants involved themselves with hygiene problems in current OP applications with surgical robots ("as-is" analysis) and the evaluation of future ones (trend prognosis). This process lasted a half-day and resulted in the workshop's search field being divided systematically into unique and clear subfields. From these four concrete fields, the participants chose those that needed to be solved with the highest priority. In order to develop these further, four subgroups were formed consisting of five to six participants and a moderator. In the remaining one and a half days, the group worked intensively on developing innovative solutions and in doing so produced numerous sketches and simple models from a variety of materials. The makeup of the groups varied throughout depending upon the interest of the participants. The work of subgroups was shared between all groups via presentations and discussions.

At the end of the second workshop day, four complete and detailed concepts developed from the numerous individual ideas and solutions. One was the development of a new type of foil to cover the surgical robot that could remove the current sterility and handling problems. The second subgroup suggested a total solution for the stifling of the mist that is produced during the operations (so-called aerosols, moisture of water as well as bone, and blood particles). The third solution consisted of a new, integrated system for the sterile storage of the patient's leg during a hip replacement surgery. The forth group developed a process for the cleaning of the surgical robot after the completion of the procedure.

In a subsequent evaluation of the developed concepts by the participants of the workshop, the results of the first three groups performed outstandingly with respect to the criteria of originality, problem solving capacity, and its ability to be realized. It should not be forgotten however that the participants of the workshop were made up of leading surgeons, OP nurses, engineers, and hygiene experts. The evaluation of the ideas was based on a broad as well as a deep knowledge in the field of robot-assisted surgery.

The workshop provided J&J with the beginnings for the development of new product lines. All four of the concepts included products that were not currently in the J&J product program. Some of the product ideas are still not available from any manufacturer in the market, making them world firsts.

Discussion

The deliberate alignment with the customer and the desire to make compatible, significant, and innovative jumps belong to the greatest challenges of innovation management. Companies that are able to do both increase the probability that their new products and services will perform successfully in the market. As shown here, leading companies depend upon the intensive cooperation with Lead Users who are ahead of their time in both their needs and demands from products.

The effectiveness of the Lead User methodology has been proved in some first applications. A investigation within 3M, who have great experience with Lead Users to date, supports this emphatically. In a comparison between Lead User projects and traditionally carried out development processes, the degree of innovation, turnover, market share, ability to be realized, and strategic importance were all measured as being significantly superior in the Lead User projects. From these Lead User projects, new product lines resulted that without exception were responsible for a large proportion of the turnover. It has been shown that concepts developed with Lead Users have, on average, the potential to produce turnovers eight times higher than ideas produced via traditional innovation projects (Morrison et al., 2001; Lilien et al., 2002).

The Lead User application at J&J described above resulted not just in the possibility for the further development of the existing market offer but also the chance for the expansion of their program as well as the development of completely new product lines. Beginning with conventional patient coverings and OP garments, a mature field with few possibilities to develop decisive competitive advantage, the starting points for significant innovations were developed.

Comparing one of the 3M and the J&J Lead User application reveals another interesting aspect. Even when companies start a Lead User project in more or less the same product arena ("Surgical drapes and related hygiene products"), the chances that they will end up with the same ideas for radical innovations are very limited. Why? The direction of the search fields is closely linked to the trends a company identifies and believes in to be most relevant. Since very often a high number of trends exist for a given product field, the chances that competitors would choose one and the same trend are at least not very likely. The key question for a company applying the Lead User approach therefore is, what is a relevant trend to fully concentrate on for the last two phases of the Lead User project.

In addition, with all the positive results, it should not be forgotten that the Lead User approach is no substitute for proven methods of innovation management and market research. It leads companies primarily to new ways to develop innovative products and services. As a result it produces "market-fit" ideas for innovations that must be advanced further with traditional methods of market research and engineering for the development and evaluation of concepts and prototypes.

Managerial implications

- Leading companies depend more and more on an intensive cooperation with Lead Users, individuals or organizations that are ahead of their time with regard to both needs and demands for future solutions (products and services).
- Lead Users can systematically be identified by companies, willing to open up for a close and trustful cooperation with market partners ("Open Innovation"). The five step phase approach presented in this paper has already proven useful in a number of projects in different product contexts (consumer as well as industrial product settings).
- Concepts developed with and by lead users have, on average, a higher "breakthrough" potential compared to concepts generated by traditional market research approaches.
- In our work together with J&J, presented here, we could further demonstrate that closely working together with lead users can lead to incremental product improvements and product line extensions, as well. Therefore, working together with lead users supports development teams to create *different* qualities of innovation (from small, incremental to large, breakthrough innovations), depending on the scope and objectives of the project.
- Our work further shows that large steps forward (*breakthroughs*) are more likely if the teams expand their search for lead users to analogous search fields.

References

Franke, N. and Shah, S. (2002). "How Communities Support Innovative Activities: An Exploration of Assistance and Sharing among End-users." *Research Policy* 32(1), (January 2003): 157–78.

Herstatt, C. (1994). "Realisierung der Kundennähe in der Innovationspraxis." In T. Tomczak and C. Belz (Eds), *Kundennähe realisieren*. St. Gallen: Verlag Thexis: 291–307.

Herstatt, C. (2002). "Search Fields for Radical Innovations Involving Innovation." *International Journal of Entrepreneurship and Innovation Management* 2(1): 473–84.

Herstatt, C. and Lüthje, C. (2003). "The Lead User Method. Theoretical-Empirical Foundation, Key Issues for Research and Practical Implementation." R+D-Management 2003 Conference Proceedings. Manchester (UK).

Lilien, G. L., Morrison, P. D., Searls, K., Sonnack, M., and Hippel, E. von (2002). "Performance Assessment of the Lead User Idea Generation Process." *Management Science* 48(8): 1042–59.

Lüthje, C. (2000). *Kundenorientierung im Innovationsprozess. Eine Untersuchung zur Customers-Hersteller-Interaktion auf Konsumgütermärkten.* Wiesbaden: Gabler.

Lüthje, C., Herstatt, C., and Hippel, E. von (2002). *The Dominant Role of "Local" Information in User Innovation. The Case of Mountain Biking.* Working Paper: Massachusetts Institute of Technology.

Morrison, Pamela D., Lilien, Gary L., Searls, Kathleen, Sonnack, Mary, and Hippel, E. von (2001). *Performance Assessment of the Lead User Idea Generation Process for New Product Design and Development.* Working Paper: Massachusetts Institute of Technology.

Morrison, Pamela D., Roberts, John H., von Hippel, E. (2000). "Determinants of User Innovation and Innovation Sharing in a Local Market." *Management Science* 46(12) (December 2000): 1513–27.

Shah, Sonali (2000). *Sources and Patterns of Innovation in An Consumer Products Field. Innovations in Sporting Equipment.* Working Paper: Massachusetts Institute of Technology.

Shaw, Brian (1985). "The Role of the Interaction between the User and the Manufacturer in Medical Equipment Innovation." *R&D Management* 15(4): 283–92.

Tietz, R. (2003). "The Significance of User Innovations – An Assessment from a Technological and Market Perspective." Unpublished Masterthesis: Technical University of Hamburg.

Urban, Glen L. and von Hippel, E. (1988). "Lead User Analyses for the Development of New Industrial Products." *Management Science* 34(5) (May 1988): 569–82.

von Hippel, E. (1977). "The Dominant Role of the User's in Semiconductor and Electronic Subassembly Process Innovation." *IEEE Transactions on Engineering Management* 24(2): 60–71.

von Hippel, E. (1988). *The Sources of Innovation.* New York: Oxford University Press.

11

Relying on Experts: How to Effectively Gather Information for Innovation Projects from Market Specialists

Cornelius Herstatt, Christian Lüthje, and Christopher Lettl

Introduction

The ability to successfully develop and market innovative new products is critical for corporate growth and sustainable competitive advantage. For innovations projects, firms need to collect, process, and filter a substantial amount of information: innovation management is therefore to a high degree information management. Particularly in the early phases of the innovation process, the so called "fuzzy front end," firms need to develop an understanding of various topics: What are current and future customer needs? Which trends and technological developments are going to have an impact on the business? What might be promising avenues for new solutions? Information with regard to these questions is essential in the early phases of the innovation process since they provide hints for ideas and concepts that are both, customer and future focused (Khurana and Rosenthal, 1997; Reinertsen, 1999). Due to the fact that most of this information is located externally rather than internally, secondary sources of information need to be screened intensively. Besides the internet, written documents and electronic databases, external experts appear to be a key source of information in the fuzzy front end. An expert is someone who can perform, in a specific domain, at a level that exceeds ordinary individuals (Lunce et al., 1993). The advantage of experts as information providers is twofold: Firs, experts can deliver state-of-the-art information about the field of interest. Second, by interviewing experts, the type and amount of information gathered can be matched to specific information needs (Arken, 2002).

When planning to incorporate experts during the early phases of innovation projects, firms are confronted with the question: how is knowledge distributed among relevant experts? Do all experts share the same knowledge about a particular field of innovation? Or does each expert develop a unique knowledge base that differs substantially from those of his expert colleagues? In other words: is knowledge distributed homogeneously or heterogeneously among a given pool of experts?

The type of knowledge distribution has direct implications for the involvement of experts in innovation projects. In particular, knowledge distribution impacts on the efficiency of the information gathering process. Due to time and budget restrictions, firms need to perform a thorough analysis of how many experts need to be interviewed to develop a comprehensive and current understanding of the search field. In the case of a homogeneous distribution of knowledge we assume that any expert provides similar information. Therefore, only a limited number of experts need to be interviewed. If knowledge were allocated heterogeneously however, the chances are high that any expert would provide distinct information. This implies that innovation teams might be forced to interview a larger group of experts.

To shed light on the question of how knowledge is distributed among experts we conducted an empirical study in the field of surgical hygiene products. Market experts in this product area were interviewed in order to analyze the type of information that could be provided by different experts.

We found that knowledge is distributed rather heterogeneously. The information provided, was very seldomly shared by more than one expert. Experts appear to focus on particular knowledge corridors that are not necessarily congruent with the knowledge field of others. The findings suggest that this is mainly due to the volume and complexity of information within one product field, hence forcing experts to focus on specific knowledge areas. We also found that the knowledge focus of a given expert is closely linked to his specific (professional) background and largely depends on the relationship that the expert has with the product area in question. As a result of our study, it seems risky during the early stages of the innovation process to restrict external information gathering to only a small group of experts. However, clearly the number of interviews could possibly be reduced by gathering information about relevant experts prior to starting the interviews.

This chapter is organized as follows. In the next paragraph we introduce theoretical perspectives relevant for the research question. In the third section we introduce the context and research methodology of our empirical study. Finally, we present the results of our study and discuss their implications.

Distribution of expert knowledge within a search field – theoretical framework

Are experts arbitrarily exchangeable in the information gathering process given that they have roughly the same kind of knowledge? Or is an expert unique in the way that they possess a proprietary knowledge base? There are theoretical arguments for both views.

Neoclassical theory of perfect competition is based on the assumption of perfect information. Under this constellation any agent within a market system has access to absolutely all relevant information that is needed to make rational decisions. In other words: neoclassical economists assume complete information transparency. Therefore the market system is characterized by information symmetries, as every agent possesses the same knowledge and information base (Demsetz, 1997): in the neoclassical framework knowledge is distributed homogeneously among the agents of the market system (Deligönül and Cavusgil, 1997; Demsetz, 1997).

Advances in information and communication technologies support the idea of complete information transparency. Enabling a vast and intensive exchange of information on a global scale, these new technologies lead to a comprehensive diffusion of knowledge within the agents of a market system in a short time. Particularly individuals that work on the forefront of a certain issue can benefit from the new technologies as state-of-the-art knowledge can easily be transferred around the globe. Experts of a certain community can discuss important issues online via chat rooms; alternatively they can exchange e-mails, transfer documents, and conduct video-conferences. Therefore, new information and communication technologies contribute to the extensive transfer of information and thus to a shared knowledge base among experts (Detmer and Shortlife, 1997).

Drawing a sharp contrast to neoclassical theory, there are two theoretical frameworks that claim a heterogeneous distribution of knowledge among the agents of a market system: market process theories and new institutional economics.

The basic assumption of market process theory is that information within a market system is dispersed in a way that any agent possesses different knowledge. Market process theory therefore presumes a decentralization of knowledge within a society (Hayek, 1945; Ikeda, 1990). If knowledge is allocated differently among the agents of a market system, we would expect that experts, even within the same search field, develop a proprietary knowledge base that differs from the know-how of other experts within the same field.

Similar to market process theories, also new institutional economics abandon the neoclassical assumption of perfect information (Palermo, 1999). This becomes evident in the key assumption of new institutional economics that the rationality of individuals is not perfect but rather "bounded." The concept of "bounded rationality," which was first introduced by Simon challenges the neoclassical concept of the homo economicus. While the latter acts under perfect rationality (full knowledge of goals, alternatives, and outcomes) and is thus able to identify the optimal decision in a given environment. Simon's decision maker is restricted in terms of mental capacity (Simon, 1957). The expression "bounded rationality" is used to denote a type of rationality that people resort to when the environment in which they operate is too complex relative to their limited mental capabilities (Dequech, 2001). According to Simon's concept, agents are not able to gather and process absolutely all of the accessible information within a market system. Every agent perceives only a limited but unique set of information. Consequently, information and knowledge asymmetries emerge among the agents of a market system. Experts of a certain domain might need to focus their information gathering and processing activities on a fraction of the entire search field due to limited cognitive receptivity. As individuals are not able to perceive their domain as a whole, they focus on the specific context in which they operate. Information processing and knowledge development therefore can not be separated from the context. According to Fleck, the context is defined by three elements:

Domains: (the more or less defined areas or "parts of the world" to which the particular expertise applies. Thus the domain is similar to the search field of an innovation project.)
Situations: (components, people, domains, and other elements present at any particular instant of expert activity.)
Milieux: (essentially the immediate environments in which expertise is exercised: comprising set of situations occurring regularly at particular locations, for example laboratories, offices.)

As the development of knowledge is always embedded into a specific domain, a specific situation, and a specific milieux, the expertise of individuals is never quite the same. This holds true even for the knowledge development within a specific domain as the distinct situation and milieux lead to the development of a unique knowledge base (Fleck, 1997). Therefore, "bounded rationality" together with its implied focus on the specific context suggests the development of specialized know-how and thus a heterogeneous distribution of expert knowledge. This assumption is supported by a recent study of Franke/Hippel. They focus their analysis on users of opensource software who can be seen as experts for market needs. The findings reveal that these users develop highly specialized needs due to their specific use context (Franke and von Hippel, 2002).

There is, however, a counterargument about the implications of "bounded rationality" on the knowledge distribution among experts. As experts can only absorb a limited amount of information they might focus their information gathering and processing activities on only the most significant information. Examples of such significant information are the most urgent problems, the most significant trends and the topics with the highest priority in the search field. If every expert were able to identify this highly relevant information, we could expect a high consensus among experts regarding the information they would provide in interviews. Consequently, "bounded rationality" could also imply that knowledge about the most important issues is homogeneously distributed among experts. There is empirical evidence that supports that argument. In their study Griffin/Hauser address the question of how many customers are to be interviewed to identify the customer needs in a particular market. Results of their study reveal that the interviewing of relatively few customers is sufficient to achieve a comprehensive understanding of current and presumably the most important customer needs (Griffin and Hauser, 1993). This implies that customers share a common understanding about the most important topics related to a given product.

To sum up, there is theoretical and empirical support for both types of knowledge distribution. While neoclassical theory of perfect competition implies a homogeneous allocation of knowledge among experts, market process theories and new institutional economics propose a heterogeneous distribution. The present study therefore aims to develop insights into knowledge distribution in the specific context of external information search for innovation projects. To explore which types of knowledge distribution can be found among external market experts we conducted an empirical study in the field of medical products. The research context and methods are described in the following section.

Research context and methods

Research context: Lead User study in the field of medical hygiene products

The topic of this research is the distribution of knowledge among experts that are to be used as an external source of information during the early stages of innovation projects. We explore this question in the context of an application of the so called "lead user method" (von Hippel, 1988).

Lead users are advanced product users who are both motivated and qualified to make significant contributions to the development of radically new products. Lead users' qualifications evolve from the fact that they anticipate the future needs of the market and do so significantly earlier than the majority of other customers. Lead Users therefore benefit significantly from innovations that meet their emerging needs. Manufacturers can use lead users as "need forecast laboratory" for emerging markets (von Hippel, 1986).

The identification of advanced customers is supported by the lead user method. The methodology consists of a multiphase process aiming to develop product ideas and concepts by integrating lead users into a workshop. Since it is an axiom that lead users are one step ahead of critical issues that will impact future markets, the process starts with determining those issues. To find out, innovation project teams try to identify urgent problems, ideas for solutions, and, most importantly, critical trends in the particular search field. Interviewing experts has proven to be especially valuable for gathering this type of external information. Thus, lead user teams usually talk to people in the field who have a broad view of emerging trends and are leading edge with respect to the topic being studied (Herstatt and von Hippel, 1992).

This stage of the lead user method – the interviewing of experts to get external information for idea generation – is the context of the present survey. The information gathered in the course of expert interviews forms the basis of our analysis on how information is distributed among experts.

The focus of our empirical work in the medical market is the research field of surgical hygiene products. This product category encompass all disposable articles aimed to prevent infections of the patient and the medical staff during surgery, such as surgical drapes, gowns, masks, and other garments. The lead user project had the objective of finding advanced users in order to develop completely new products that would capture significant market share in existing markets or would create totally new markets.

Expert interviews

In order to determine actual problems, ideas for solving problems and future trends in the field of surgical infection control, interviews with experts from three areas were held:

Normal users: The first expert group consisted of users from the target market that is from surgeons and surgical nurses. Those questioned included doctors who were leaders in their fields and had implemented new technology in surgery.

Extreme users: Also included were "extreme users." This group consisted of doctors who worked under particularly difficult conditions such as in the tropics (poor hygiene controls, poor product availability) or burns surgeons (long operations with large, moist wounds). In contrast to their counterparts in the "normal user" group, they are confronted with extreme challenges to prevent infection of their patients.

Hygiene experts: In addition to the actual product users, experts in the subject of clinic hygiene were also interviewed. Hospitals have to employ medical staff in charge of ensuring that hygienic standards are met in all areas of clinics and hospitals. Frequently, hygiene experts have their academic background in bacteriology and virology.

The experts were identified by screening different sources of information like the internet (for example web pages of leading clinics, research institutes, and medical associations) and magazine publications. We contacted the identified experts and, were able to carry out interviews with 12 normal users, 6 extreme users, and 6 hygiene experts, making 24 interviews in total. The chronological order of interviews was not consciously determined by the authors, but was mainly determined by finding the earliest opportunity to fix an appointment to interview the experts.

Subject to geographical spread, we conducted either one-to-one or telephone interviews. To ensure a consistent interview structure, open questions about problems, solutions, and trends in the search field were fixed in an interview guideline. However, the interviewers had the opportunity to ask further questions in reaction to the remarks of the respondents (semi-structured). All interviews were recorded in writing.

Coding and evaluation of the information

To analyze the information provided in the interviews, we performed a content analysis of the interview protocols (Berelson, 1971; Krippendorf, 1980). In the context of the present survey, the expert interviews aimed to identify hints for new product development and lead users who are ahead of the market. We therefore distinguished three categories of information:

Actual usage problems,
Solution ideas for usage problems, and
Trends having an impact on product use.

Each information category was specified by an operational definition. In a preliminary analysis of a sample of interviews, this category system proved to be free of overlaps and exhaustive to categorize the information provided in the protocols. The unit of analysis was one remark which provided a complete description of a problem, solution or trend. In the main, one information unit encompassed from one to three sentences from the interviewees.

The information units within the protocols were screened and coded by one author of this article. Four additional coders went through a random sample of ten of our interviews and were asked to identify the information units in the protocols. The four coders were members of the research team and were therefore familiar with the search field of infection prevention. To test reliability between the coding of the first author and the other four coders, we calculated the Cohen's Kappa coefficient for each of the three information categories (Cohen, 1960). This measure takes into account that a certain percentage of corresponding assignments can already be expected in a random coding. Thus, this coefficient is stricter than the simple pair wise inter-coder reliability. Cohen's Kappa was on average 84.9 percent for all three categories (0.86 for usage problems, 0.88 for solutions, and 0.72 for trends). This is a very satisfactory value that indicates a high reliability of the coding procedure. Thus, the coding of the first author formed the basis for the analysis of information heterogeneity.

In order to be able to prioritize the experts' remarks, the information units were rated according to their relevance. Relevance in the context of the present innovation area stands for the impact of the particular problem, solution or trend regarding the hygiene situation in the operation room, for example: (1) Problems with direct and serious consequences for wound infection, (2) solutions that would directly result in major improvements in infection prevention, and (3) trends with clear and substantial future impact on hygiene are to be rated as the most relevant information units. Since a five-point-rating scale proved to put too big a strain on the discriminatory ability of the rating people, we made use of a three-point-rating scale (high, middle, and low importance). Again, all information units provided were initially rated by one of the authors. The same four coders mentioned above rated ten randomly selected information units. On average 77.5 percent of the ratings matched between the author and the four other rating agents. When the rating was not consistent, the difference was never more than one scale point. In essence, the reliability of the relevance rating was satisfactory.

Findings and future research

Distribution of information between different expert groups

A pattern of heterogeneous information distribution among the three expert groups "normal users," "extreme users," and "hygiene experts" emerged. In

Table 11.1   A Small fraction of total information is shared by more than one expert group

Type of information unit	Fraction of total information units that was revealed by ...			
	...One group	...Two groups	...Three groups	Total
Notions of problems	84.7% (50)	11.9% (7)	3.4% (2)	100% (59)
Notions of solutions	95.0% (19)	0% (0)	5.0% (1)	100% (20)
Notions of trends	87.5% (14)	12.5% (2)	0% (0)	100% (16)
Total	87.4% (83)	9.5% (9)	3.1% (3)	100% (95)

Numbers in brackets stand for the absolute number of information units.

Table 11.1 it is shown whether the information that resulted from all expert interviews was revealed by only one, by two, or by all three of the expert groups.

The findings indicate that for all types of information, most units of information were provided by one expert group only (87.4 percent). The problems, solutions, and trends mentioned by one expert group do not usually correspond with the notions of the experts in the two other groups. Consequently, a particular piece of information is rarely mentioned by members of two (9.5 percent) or even all three groups (3.1 percent). It appears that different expert groups do not have access to the same information and do not share a common understanding/opinion about the issues affecting a particular product field.

However, the heterogeneity of knowledge seems to decrease when focusing on the most relevant and important information. As mentioned above, problems, solutions, and trends were rated on a three-point-rating scale (high, middle, low relevance) with respect to their impact on infection control. The results in Table 11.2 indicate, that the information units that were provided by more than one expert group are, on average, more relevant than the knowledge held exclusively by one of the three groups of interviewees (1.93 versus 2.34; 1 = high importance, 3 = low importance). In other words, the knowledge that is distributed throughout many different types of experts tends to be more important. However, this does not imply that every piece of significant information on a given subject could be provided by any expert somehow related to a particular search field. The most information units that were rated as being highly relevant for infection control is still exclusively mentioned by the members of one expert group (75 percent of all highly relevant information).

Strong indication exists that the information held by an expert is linked to his specific relationship to the product category in question. Again, this becomes obvious by comparing the three expert groups; "normal users," "extreme users," and "hygiene experts" (see Table 11.3).

Table 11.2 Shared information is on average more relevant

	Information revealed by one expert group only (n = 87)	Information revealed by two or three expert groups (n = 14)	Sig.[b]
Relevance (mean)[a]	2.34	1.93	p < 0.05

[a] Relevance was measured using a three-point-rating scale (1 = high, 2 = middle, 3 = low relevance); n = 95.
[b] Two-tailed t-test for independent samples.

Table 11.3 The expert groups emphasize on different types of information

Type of expert group	Type of information unit			
	Notions of problems	Notions of solutions	Notions of trends	Total
Normal users	64.4% (38)	28.8% (17)	6.8% (4)	100% (59)
Extreme users	80.0% (12)	20.0% (3)	0% (0)	100% (15)
Hygiene experts	58.3% (21)	5.6% (2)	36.1% (13)	100% (36)
Total	64.5% (71)	20.0% (22)	15.5% (17)	100% (110)

Chi-Square = 21.19; df = 4; p<0.001; numbers in brackets stand for the absolute number of information units.

When the information units are segregated into problems, solutions, and trends, we find that "normal users" primarily mention problems that they face when using the products. To a smaller extent they suggest solutions for problems and even more seldom they mentioned trends in the search field (see first row of Table 11.3).

"Extreme users" showed a much stronger emphasis on usage problems. More than three out of four remarks made by members of this expert group were centered around a problem in product usage. Accordingly, the percentage of solutions and trends is much smaller than in the group of "normal users" (second row of Table 11.3). The difference is probably due to the much more challenging usage situations that the extreme users are confronted with. This group of doctors works under particularly difficult conditions (for example in the tropics) or have to treat extremely difficult cases (long operations with large, moist wounds). In this context the probability of problems in the application of products for infection prevention increases. Consider for example the adhesive strip that is necessary for attaching the surgical drapes to the patient. In most surgeries the existing products offered by the medical industry can guarantee reliable adhesion and, at the same time, avoid harming the patients' skin. Both requirements are much more difficult to ensure in the case of burn surgery since long operations with moist wounds and the application of large amounts of cooling water endanger the adhesion of the stripes. In addition, large areas of burned skin hamper a risk of less adhesion of the drapes to the skin of the patient.

The "hygiene experts" were able to indicate more trends than the respondents in the other two groups (third row of Table 11.3). They seem to focus more on the general context of product use and are able to take a forward thinking perspective on the broader field of hospital hygiene. Surgical hygiene products such as drapes, masks, and gowns are only one aspect of hygiene-related issues. The experts in this group seem to perceive a certain amount of problems in product usage as they endanger infection prevention. However, they seldom thought about specific solutions for these problems probably because they do not use the hygiene products themselves. Instead, they are in a much better position to report about developments and trends that might have an impact on the future demands on surgical hygiene products. For instance, changes in the norms and standards for hospital hygiene are an important issue that clearly falls into the responsibility of hygiene experts, but are hardly known by the nurses and surgeons as the product users. Not surprisingly, this "trend" information was mostly provided by the hygiene experts.

Another indication for a strong association of the information provided by experts and their usage context is presented in Figure 11.1. Again, this analysis differentiates the information according to their relevance for infection control on a three-point-rating scale (high, middle, low relevance). Extreme users mentioned more problems, solutions or trends with a high and direct impact on infection control than their counterparts in "normal" surgery fields. Similarly to the line of reasoning above, this might be explained by the more challenging product usage by the doctors working in fields like tropical and burn surgery. The problems that they encounter seem to be more severe and consequently their suggested solutions have a clearer and

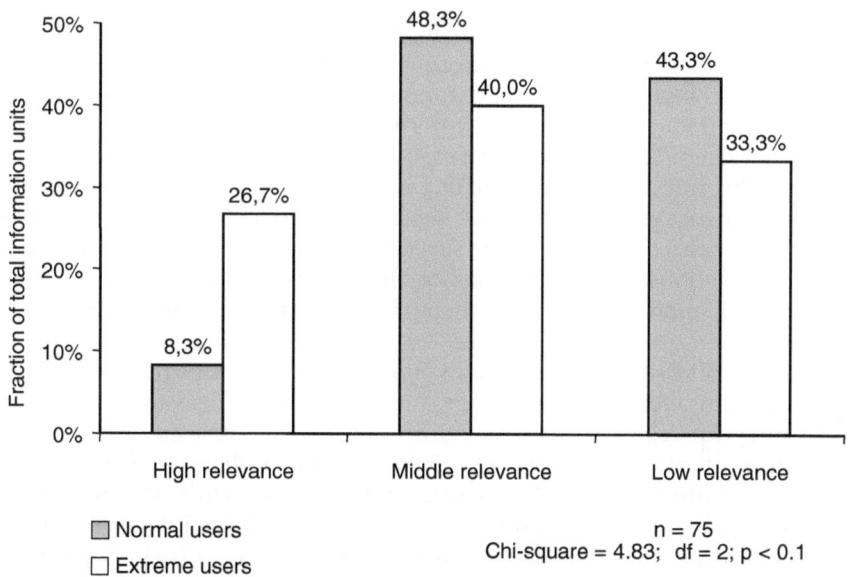

Figure 11.1 Information provided by extreme users tends to be more relevant for infection control than remarks from the normal user group

more direct impact on the prevention of infections during the course of operations.

In total, the type and relevance of information provided by the respondents is determined by the particular relationship that an expert group has to the product field. The different experiences and usage of the experts explains why knowledge and information is rather heterogeneous. Expert groups stress different topics because they deal with the products in a different way. However, with respect to the most important information, the knowledge is slightly more homogenous. The relevance of information tends to be somehow associated with its distribution.

Distribution of information within expert groups

In summary, the findings in Section 5.1 suggest that knowledge about problems, solutions, and trends in a particular product field is in the main not shared by different expert groups. Of course, one could argue that while this is comprehensible when analyzing different types of experts, this might not be the case when comparing the interview outcomes within a given expert group.

To test this possibility we analyzed the redundancy of information within the three expert groups "normal users," "extreme users," and "hygiene experts." More precisely, we explored how many additional information units were provided by each interview when compared with the interviews that have been conducted in the same group before. For this analysis, it was necessary to bring the interviews within each expert group in an interview sequence. Since the amount of information varied significantly between the experts of the same group, the chronological order of the interviews influences the amount of additional information gained during the course of the interviews. For instance, if the sequence started with the most informative talks, information redundancy would be comparatively high in the interviews that followed and vice versa. The particular order of interviews that was realized in the present survey is incidentally and mainly due to time restrictions and availability of the experts we contacted. It seems therefore inappropriate to base the analysis on this particular sequence of interviews.

In the following analysis we assume a setting where no preliminary information about the knowledge and the cutting edge status of potential interviewees exists. We therefore also assume that any sequence of interviews within an expert group has the same probability to occur. The interviews are interpreted as a finite set of events with an equally distributed probability. Consequently, a particular order of interviews can be seen as one permutation that is equally probable as any other permutation of interviews.[1]

The analysis was based on six talks in each group leading to 6! different permutations for the interview sequence in each expert group. Since this is a large number, we randomly simulated 700 orderings. For each permutation, the amount of nonredundant information units gained in interview to interview was calculated and averaged over all permutations. The results in Figure 11.2 stand for the expected net information gain per interview as a percentage of all information gathered in the six interviews – the percentages within each group therefore add up to 100 percent. It is important to note, that the findings are based on the assumption that no preliminary selection of the experts is possible.

As indicated in Figure 11.2 the trend seen in the graphs is similar for all three groups of experts: The number of new information units decreases during the course of the interviews. Interviewing more experts leads to more useful information. However, the decline is not substantial. More than 40 percent of all information is provided within the second half of the sequence of interviews. At least 10 percent of the total information can still be expected in the last interview.

To identify most of the problems, solutions, and trends that are known in a particular expert group, it is not sufficient to interview one or two experts. Even the experts in the same category don't seem to share the same core of knowledge. Instead, knowledge seems to be heterogeneously distributed due to different personal experiences of the experts – even if they are experts in the same field.

This finding is supported by another analysis of information redundancy. We investigated how many information units exclusively one expert within a group mentioned and what fraction of total information was jointly

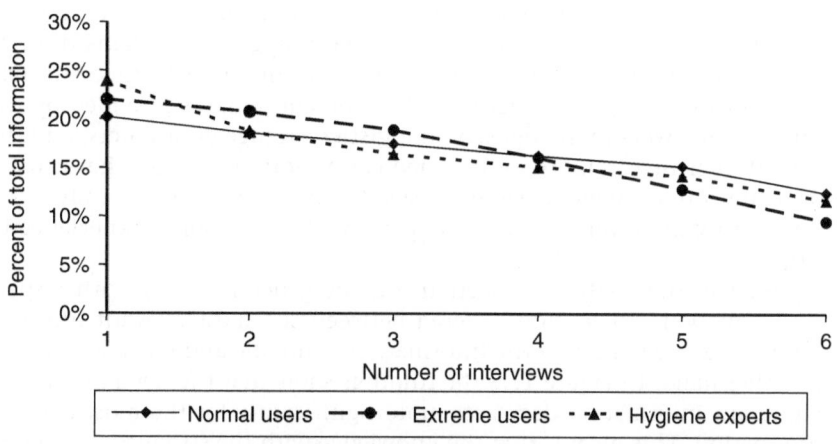

Figure 11.2 In a random order of interviews a significant fraction of total information is still provided during the last interviews

Table 11.4   The majority of information was not shared by more than one expert

	Fraction of total information that was revealed in altogether...						
	1 interview	**2** interviews	**3** interviews	**4** interviews	**5** interviews	**6** interviews	**Total**
Normal users	78.3% (47)	11.7% (7)	6.7% (4)	1.7% (1)	1.7% (1)	0% (0)	100% (60)
Extreme users	66.7% (10)	26.7% (4)	6.6% (1)	0% (0)	0% (0)	0% (0)	100% (15)
Hygiene experts	80.6% (29)	8.3% (3)	5.6% (2)	2.8% (1)	2.8% (1)	0% (0)	100% (36)
Total	77.5% (86)	12.6% (14)	6.3% (7)	1.8% (2)	1.8% (2)	0% (0)	100% (111)

Chi-Square = 4.17; df =10; n.s.; numbers in brackets stand for the absolute number of information units.

revealed in two, three, and more interviews (see Table 11.4). These percentages do not vary significantly among the three different groups of experts. It becomes obvious, that in all three groups a large fraction of the total information was exclusively provided by one interviewee (77.5 percent). Only 12.6 percent of all information units were shared by two interviewees and even less were mentioned by three (6.3 percent), four (1.8 percent) or five experts (1.8 percent). No single information unit was mentioned by all six interviewees within a particular group of experts.

To illustrate the link between the particular use context and personal experience of any expert and his information input, consider the case of one member of the expert group "hygiene experts" who had been working for many years as a surgery nurse before taking on the task of hygiene agent in a hospital. This person noted many hygiene problems directly associated with the use of hygiene products in surgery since she had made personal usage experiences. The problem remarks provided by the other experts were more focused on logistics, storage of products, and the general hygiene discipline of the medical personnel – issues that usually form the core of hygiene agents' tasks. Problems relating to product use were hardly mentioned by these experts without personal experience in surgery.

Similar to the analysis in Section 5.1, the general finding with respect to information distribution differed between the most relevant information on the one hand and the information of middle and low relevance on the other hand. In Figure 11.3 the findings show that the net gain of non-redundant information decreases more significantly for the most relevant information. This means, that the knowledge with the strongest impact on infection prevention is, again, more widely shared among the members of an expert group than other, less important information. However, neither

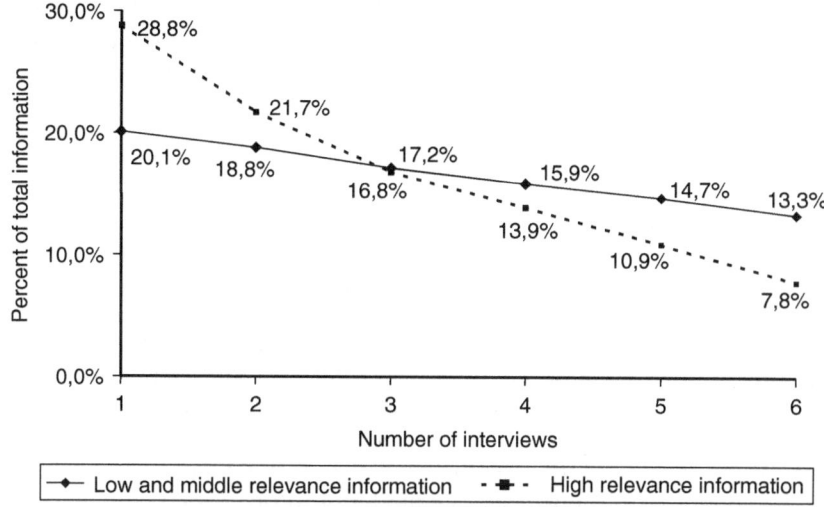

Figure 11.3 In a random interview sequence the acquisition of nonredundant information decreases more significantly with high relevance information than with the information of low and middle relevance

of the graphs shows a substantial difference in terms of percentage values. The findings therefore do not at all imply that the important knowledge can be elicited by interviewing one or two experts of the same type. Even in the last interview, very relevant remarks are made that were not mentioned by the other respondents in the interview sequence.

Conclusions

The results of the present survey show high heterogeneity of expert knowledge. Experts in our sample vary in terms of the problems identified, proposed solutions, and trends in a particular product field. This is equally valid for the group based analysis between the expert groups as well as in the individual based analysis within each expert group. In essence, the homo economicus that is acting in contexts of transparent and perfectly distributed information is not reflected by the findings.

The difference of information held by different parties can probably not be explained by a lack of information exchange between experts in the present product field. The medical community is a close knit community that is very open to the exchange of the latest knowledge via conferences, journals, internet forums, and informal networks.

This exceptional situation might initially be due to the common sense of obligation to diffuse information that might be necessary to save lives and reestablish the health of patients. In addition, by freely revealing personal knowledge, manufacturers of hygiene products can be prompted to realize product improvements that the users can benefit from Harhoff et al. (2000).

Second, the apprehension to loose proprietary knowledge – a potential barrier to the free sharing of information – seems not to be relevant for the product field in question. Hygiene products, such as surgical drapes, gowns, and masks, while indispensable for preventing infection do not improve competitive advantages in surgery. Doctors interested in pushing their field will innovate in surgical methods and complex instruments instead. Furthermore, most surgeons, due to geographical spread or different specializations, are not necessarily direct rivals in the same area. To reveal information about problems, solutions, and trends in a product field will therefore not compromise competitiveness (Harhoff et al., 2000).

If it is not the lack of information exchange, then it seems reasonable that the heterogeneity of knowledge is a matter of information complexity and volume (Hayek, 1945). The sheer volume of information that has been generated for a given product field seems to exceed the mental capacity of individuals – even the capacity of experts. Cognitive limits and specialization of knowledge inhibit that a single experts can provide all of the information relevant to a particular subject. As a consequence, in alignment with the concept of "bounded rationality," individuals gather only a limited subset of relevant information.

For our sample it was shown that the specific knowledge base of a particular type of experts can be explained by their unique use context and relationship to the search field. As proposed by the theory of contextual development of knowledge, information gathering depends on the specific context in which the expertise is built on. Extreme users, when asked, stress problems in product usage to a greater extent than their counterparts in hygienically less challenging surgery fields. Hygiene experts – generalists without a large amount of personal usage experience – mention more trends and less concrete product solutions than the experts in product use. Also, for experts of the same type, anecdotic evidence indicates that the knowledge differences are associated with the personal experiences and usage patterns. The different domains and Milieux that the people are working in and the different situations with which they are confronted, in fact, drive knowledge development.

This pattern can be understood in terms of costs and benefits of information gathering. With respect to information costs, consider that experts will often acquire information without additional effort in the course of their work activities. If an expert decides to engage in an information search outside his own scope, higher investments in time and financial resources are usually required. The costs of information gathering are particularly high in cases where the information is "sticky" (von Hippel, 1944). Information stickiness can be due to attributes of the information such as how easy it is for information seekers to encode the information. For the expert users interviewed in the survey, it seems reasonable that a considerable part of their knowledge is rather tacit, since it is acquired in the course of product use (Polanyi, 1958). Often it is a problem for customers and users to clearly articulate the problems they face when using the products and to suggest how these problems could be solved.

The same can be argued for the benefit of information. Information relating to their specific working context should have the highest personal benefit for the experts. Through the "in-house" use of this information they can directly enhance their working situation. In contrast, the anticipated reward of information that is rather relevant for others is lower. As mentioned earlier, monetary gains by licensing or selling this information are quite improbable, particularly in the search field of this survey.

Strongly connected to the issue of local knowledge variations is the question of how many experts need to be interviewed in order to identify most of the relevant information about problems, solutions, and trends in a product field. It becomes apparent that the present research has a direct practical implication for the information gathering during the early stages of product development. First, the pattern of heterogeneous knowledge among experts implies that it is risky for an innovation team to restrict external information search on one type of experts. We found, that only a small fraction of the total information was provided by experts in all three of the interviewed expert groups. Second, it also seems risky to narrow interviews to a group of similar experts on a very small number of interviewees. Our data suggest that interviews with two to three experts of a particular knowledge field should identify less than 60 percent of the information that would be obtained by asking six experts in the same key field. Even though the most relevant information on a subject is distributed slightly more homogenously and is more often shared among experts of different groups as well as among experts of the same type, the risk of missing very relevant information remains high.

At first sight these results imply that there are clear benefits for a high number of interviews. However, this implication only holds true under the assumption of no preliminary information about the amount of knowledge and information base of any particular expert. The experts in our sample, for example, vary with respect to the amount and quality of information input. To reduce the number of interviews, an innovation team could try to interview the individuals with the highest amount of knowledge as early on as possible in the process of information search. One approach is to track down especially promising experts by networking through the expert communities. In a network search the innovation project team would begin to interview one expert with apparent knowledge in the field and then ask them if they know of other experts who might be able to provide even more valuable information. The advantage of this approach is that experts tend to know others who are at the cutting edge of the subject (von Hippel et al., 1999). A team should therefore quickly identify the most attractive experts and, hence, should be able to reduce the number of interviews without loosing much of the most important information on a particular topic.

However, the networking search process is less promising in fields where it is difficult for one expert to assess the leading edge status of other experts. This might be the case if no tight networks between individual experts exist and therefore no direct interaction takes place. At the same time, the lack of close interaction would imply that the knowledge and information is more heterogeneous among experts. Therefore, it seems reasonable that an innovation team should explore the ties between experts, before deciding to take the networking approach and before assessing the number of interviews. For instance, answers to the following questions can serve as indicators for the ties between experts and the homogeneity of knowledge distribution:

Are there institutionalized platforms for a formal information exchange between experts (for example conferences, publications, chat rooms)?
Are there active informal networks activities (for example on the basis of mutual trust)?
Are there strong interdependencies between the works of different experts?
Can experts benefit from formal and informal information transfer?

Given that an expert population is rather heterogeneous with respect to the knowledge and information in stock, it seems risky to inconsiderately restrict external information search to a very small number of experts. To reduce costs and time delays in the early stages of innovation projects the project teams needs to gather preliminary information about the expert community and, if possible at reasonable costs, about individual experts. On the basis of this information it can be decided how many experts need to be interviewed and which search strategy should be followed. Moreover, our findings suggest that it is possible to predict the type of information provided by experts on the basis of the background and personal experience of the interviewee (contextual knowledge). This paves the way to efficiently gathering knowledge of any specified type if personal information about the experts is available before starting the search process.

Managerial implications

- First, the pattern of heterogeneous knowledge among experts implies that it is risky for an innovation team to restrict external information search on one type of experts.
- Even though the most relevant information on a subject is distributed slightly more homogenously and is more often shared among experts of different groups as well as among experts of the same type, the risk of missing very relevant information remains high.
- At first sight these results imply that there are clear benefits for a high number of interviews. However, this implication only holds true under the assumption of no preliminary information about the amount of knowledge and information base of any particular expert. The experts in our sample, for example, vary with respect to the amount and quality of information input.
- To reduce the number of interviews, an innovation team could try to interview the individuals with the highest amount of knowledge as early on as possible in the process of information search. One approach is to track down especially promising experts by networking through the expert communities. In a network search the innovation project team would begin to interview one expert with apparent knowledge in the field and then ask them if they know of other experts who might be able to provide even more valuable information.
- However, the networking search process is less promising in fields where it is difficult for one expert to assess the leading edge status of other experts. This might be the case if no tight networks between individual experts exist and therefore no direct interaction takes place. At the same time, the lack of close interaction would imply that the knowledge and information is more heterogeneous among experts. Therefore, it seems reasonable that an innovation team should explore the ties between experts, before deciding to take the networking approach and before assessing the number of interviews.
- Given that an expert population is rather heterogeneous with respect to the knowledge and information in stock, it seems risky to inconsiderately restrict an external information search to a very small number of experts. To reduce costs and time delays in the early stages of innovation projects the project teams needs to gather preliminary information about the expert community and, if possible at reasonable costs, about individual experts. On the basis of this information it can be decided how many experts need to be interviewed and which search strategy should be followed.
- It is often possible to predict the type of information provided by experts on the basis of the background and personal experience of the interviewee

(contextual knowledge). This paves the way to efficiently gathering knowledge of any specified type if personal information about the experts is available before starting the search process.

Note

1. We conducted 12 interviews in the group of "normal users," and respectively six interviews in the group of "extreme users," and "hygiene experts." To simplify the interpretation of the findings we standardized the data on six interviews in each group. For the "normal users" each permutation was created as a random selection of six out of 12 interviews.

References

Arken, A. (2002). "The Long Road to Customer Understanding." *Marketing Research* 14(2): 29–31.

Berelson, B. (1971). *Content Analysis in Communication Research*. New York: Hafner Publishing Company.

Cohen, J. (1960). "A Coefficient of Agreement for Nominal Scales." *Educational and Psychological Measurement* 20: 37–46.

Deligönül, S. Z. and Cavusgil, T. (1997). "Does the Comparative Advantage Theory of Competition Really Replace the Neoclassical Theory of Perfect Competition?." *Journal of Marketing* 61 (Oct): 65–73.

Demsetz, H. (1997). "The Firm in Economic Theory: A Quiet Revolution." *American Economic Review* 87(2): 426–30.

Dequech, D. (2001). "Bounded Rationality, Institutions, and Uncertainty." *Journal of Economic Issues* 35(4): 911–29.

Detmer, W. M. and Shortlife, E. H. (1997). "Using the Internet to Improve Knowledge Diffusion in Medicine." *Communications of the ACM* 40(8): 101–8.

Fleck, J. (1997). "Contingent Knowledge and Technology Development." *Technology Analysis & Strategic Management* 9(4): 383–97.

Franke, N. and Hippel, E. von (2002). *Satisfying Heterogeneous User Needs Via Innovation Toolkits: The Case of Apache Security Software*. Working Paper: MIT Sloan School of Management.

Griffin, A. and Hauser, J. R. (1993). "The Voice of the Customer." *Marketing Science* 12(1): 1–27.

Harhoff, D., Henkel, J., and Hippel, E. von (2000). *Profiting from Voluntary Information Spillovers: How Users Benefit by Freely Revealing Their Innovations*. Working Paper: Ludwig-Maximilians-Universität München.

Hayek, F. (1945). "The Use of Knowledge in Society." *American Economic Review* 35(4): 519–30.

Herstatt, C. and Hippel, E. von (1992). "From Experience: Developing New Product Concepts Via the Lead User Method: A Case Study in a 'Low Tech' Field." *Journal of Product Innovation Management* 9(3): 213–21.

Ikeda, S. (1990). "Market-Process Theory and 'Dynamic' Theories of the Market." *Southern Economic Journal* 57(1): 75–92.

Khurana, A. and Rosenthal, S. R. (1997). "Integrating the Fuzzy Front End of New Product Development." *Sloan Management Review* 38(2): 103–20.

Krippendorff, K. (1980). *Content Analysis: An Introduction to Its Methodology.* Beverly Hills: Sage.

Lunce, S. E., Iyer, R. K., Courtney, L. M., and Schkade, L. L. (1993). "Experts and Expertise: An Identification Paradox." *Industrial Management & Data Systems* 93(9): 3–9.

Palermo, G. (1999). "The Convergence of Austrian Economics and New Institutional Economics: Methodological Inconsistency and Political Motivations." *Journal of Economic Issues* 33(2): 277–86.

Polanyi, M. (1958). *Personal Knowledge: Towards a Post-Critical Philosophy.* Chicago, IL: University of Chicago Press.

Reinertsen, D. G. (1999). "Taking the Fuzziness Out of the Fuzzy Front End." *Research Technology Management* 42(6): 25–31.

Simon (1957). *Administrative Behaviour* (2nd edition). New York: Macmillan.

von Hippel, E. (1986). "Lead Users: A Source of Novel Product Concepts." *Management Science* 32(7): 791–805.

von Hippel, E. (1988). *The Sources of Innovation.* New York: Oxford University Press.

von Hippel, E. (1994). "Sticky Information and the Locus of Problem Solving: Implications for Innovation." *Management Science* 40(4): 429–39.

von Hippel, E., Thomke, S., and Sonnack, M. (1999). "Creating Breakthroughs at 3M." *Harvard Business Review* 5: 3–9.

12
Generating Innovations through Analogies – An Empirical Investigation of Knowledge Brokers

Katharina Kalogerakis, Cornelius Herstatt, and Christian Lüthje

Introduction

The phase of crucial importance in most product development projects is the front end, often called "Fuzzy Front End" (Koen et al., 2001; Kim and Wilemon, 2002). Here the decision is made which projects will get resources and which ones won't. Additionally, several studies indicate that a major part of the product-development costs heavily depend on the decisions taken in the front end (Herstatt and Verworn, 2003).

A key activity of the front-end work is to develop ideas for new products and generate product concepts. Hence the front end of product development requires creative work. A new and creative solution usually results from the combination of pieces of knowledge that have not been connected before (Geschka, 1992; Geschka et al., 1994; Hargadon, 2002). One promising avenue to create new combinations of knowledge is the use of analogies. As a basis for developing something new, one has to access one's own knowledge pool and other sources of knowledge. Accessing this knowledge and transferring it to the new solution usually requires the use of analogies – although this is not always obvious for the creative person. An analogy between two objects exists if these are similar to each other in some aspects – that is similar appearance, similar function or similar structures – and are at the same time different in some other aspects.

A key problem in using analogies for product development is to find relevant analogies early in the process. First, analogies can only be accessed if relevant knowledge of the different knowledge domains is available to the innovating person or group. Second, even if relevant knowledge is available, several factors can impede realizing the relevance of that knowledge in the current context. For example, learning is contextual, meaning that acquired knowledge is linked to the situation and meaning in which it is learned (Gick and Holyoak, 1980; Holyoak and Thagard, 1995; Schild et al., 2004).

The question arises as to which factors foster the use of analogies in product development. First, it seems to be important to have access to diverse knowledge domains. Additionally, some practice in combining knowledge from diverse sources can have a positive effect. Furthermore, open-mindedness can foster looking in different domains even though obvious similarities between the domains are rare. According to Hargadon (Hargadon, 2002, 2003) certain companies, so-called "knowledge brokers," are in a special position to use analogies better than others. These companies are familiar with a wide range of knowledge domains and are therefore able to transfer a solution from one domain to another. Their unique position in a network enables them to take advantage of so-called structural holes. Structural holes describe the separation between non-redundant contacts (Burt, 1992). For example, there is usually no contact between the individual manufacturers of sport shoes and medical devices. This structural hole was used by Design Continuum, a full-service product design firm, while developing the Reebok Pump sport shoe. They put an inflatable splint in the shoe and used a medical IV bag as the air bladder (Hargadon, 2003). This shows how bridging such a structural hole and transferring knowledge between formerly separated domains can lead to innovative solutions. Knowledge brokering can be conducted by consulting companies, design agencies, and product-development companies working for clients in diverse industries.

So far, the role of knowledge brokers in the context of product development has been explored based on a limited number of case studies (Hargadon and Sutton, 1997; Hargadon, 2002). With this research we want to enlarge this empirical basis and develop a richer understanding of the processes and the actors behind the use of analogies in product development. Specifically the present study aims to answer the following questions:

Which purpose does the use of analogies serve in product-development projects?
How do knowledge brokers ensure the access to diverse knowledge domains in order to find analogous solution approaches?

Our results indicate that analogies fulfil a greater variety of functions than usually considered in the relevant literature. For instance, analogies are not exclusively used to develop truly new products or solution strategies, but are also an important means to increase the efficiency of the innovation process. Besides, our results show that the combination of knowledge from diverse sources is primarily based on existing experience and knowledge of the persons participating in such projects. Therefore, the team configuration is a key (limiting) factor for companies tackling the development of new products, services or processes based on the usage of analogies. Hence, human resource management plays a crucial role for the formation of such product-development teams.

The paper is organized as follows: In the next section we describe our empirical approach. Afterwards we present the results of our qualitative research based on 13 cases. First, all cases are categorized according to the primary purpose to use analogies. Second, the process of using analogies is thoroughly described and analyzed. The paper ends with a summary and an outline of aspects for future research.

Empirical approach

We concentrated our research on companies that are in the position to act as knowledge brokers in product- development bridging "otherwise disconnected worlds" (Hargadon, 2002, 2003).

The companies participating in our study all offer services in product development – particularly industrial design and engineering – to clients from diverse industries. Therefore, these companies have the opportunity to transfer solutions between knowledge domains to generate innovations. In addition, they are all experts in product innovation, simply because it is their main activity.

As a research method we chose semi-structured interviews that were conducted mainly via telephone. The interview guideline was addressed to managers in charge of product-development projects in the companies participating in our research. The interviews started with a general part including questions concerning the company (for example services offered, client industries) and the interviewee (for example experience with the application of analogies in former projects). The major part of each interview focused on a project in which analogies played a role for developing a new product. The interviewee was asked to first describe this project. After this and the explanation of the analogies that were used, the rest of the questions dealt with the formation of the team and the way the team worked with the analogies. Our interview ended with an evaluation of the effects on the project resulting from the use of analogies.

Thirteen companies participated in our research, including three companies offering engineering services, four companies offering engineering and design services, and six companies that specialized solely in industrial design. The companies range from large international companies like IDEO to small companies with six-eight employees. Of the 13 persons interviewed, eight persons have a degree in industrial design (one of them additionally has a degree in engineering), four persons have a degree in engineering, and one person has a degree in IT. The projects described in the interviews can

Table 12.1 Overview of interviews

No	Knowledge broker (employees)	Industry of client	Target of project	Team configuration
1	Small industrial design studio (<10)	Music industry	Design of a mixer for audio engineering	Industrial designers
2	Small industrial design studio (<10)	Tools	Design of a steering lawn mower	Industrial designers
3	Small industrial design studio (<10)	Tools	Design study of a mini cordless electric screwdriver	Industrial designers
4	Medium industrial design studio (>30)	Vehicles	Design study of a forklift truck	Industrial designers
5	Large engineering company (180)	Medical technology	Electronics for a medical device to create high voltage in very small dimensions	Electronic and mechanical engineers, a physicist, and a technician for precision mechanics
6	Small industrial design studio (<10)	Aircraft	Design of a cockpit of a big passenger aircraft	Industrial designers and an engineer
7	Large industrial design company also offering engineering services (~ 400)	Sports	"Baton" for the Commonwealth Games	Industrial designers, human factor specialists, and engineers
8	Medium engineering company (30)	Mobile phone services	Mobile phone appliance for tunnels	A hardware engineer, a software engineer, a design engineer, and a model maker

Continued

Table 12.1 Continued

No	Knowledge broker (employees)	Industry of client	Target of project	Team configuration
9	Medium engineering company also offering design services (40–50)	"Baby equipment"	High-quality baby stroller	Engineers and designers
10	Medium Design and engineering company (45)	Medical technology	Purifier for dental tools (for example bur, polisher)	An industrial designer, mechanical engineers and technicians, a model maker and a tool and die maker, electronics and software engineers
11	Small to medium design studio (12)	Promotion	Original promotion item with long lasting value	Industrial designer
12	Medium to large industrial design studio also offering engineering services (>60)	Office furniture	High-quality backrest of an office chair	Industrial designers and design engineers
13	Large engineering company (>200)	Transport sector	Facility to rearrange allocation of train wagons	Soft and hardware engineers, technical experts

be arranged between solely technical problems and problems where design aspects dominated. An overview of the cases is given in Table 12.1.

Functions of analogies in new product development

The function of analogies in product-development processes we observed can be characterized along an efficiency / innovativeness dimension. That is, in some projects analogies are primarily used to increase project efficiency and in other projects analogies rather serve to generate highly innovative new product ideas and concepts. Also, in some projects, both strategies were followed.

The motivation to use an analogy in a product-development project has an impact on the search space explored and the type of transfer. Aiming at efficiency means a solution that fulfils the given demands in the shortest possible time – restricting the search space to well known and proximate solutions. Due to intense competition, innovating companies can be forced to use analogies to reach the given time and cost limits. How pressure from outside concerning cost and time limits can provoke the efficient use of analogies for highly innovative products is described by Majchrzak et al. in their study concerning knowledge reuse in NASA projects (Majchrzak et al., 2004).

Contrastingly, aiming at a high degree of innovation leads the team to a wide search space and a focus on far analogies. In the literature a distinction is made between near and far analogies (or intradomain and interdomain analogies respectively) (Dahl and Moreau, 2002; Bonnardel and Marmèche, 2004). If source and target of an analogy are closely related or stem from the same domain respectively, we talk about near analogies. Far analogies, on the other hand, have fewer surface similarities – here source and target belong to different domains. Dahl and Moreau have shown that the number of far analogies that are discussed during a creative product design task is an indicator for the originality of the resulting design and the appreciation of the design by the customer (Dahl and Moreau, 2002). An example of a breakthrough product innovation based on a far analogy is the Speedo Fastskin swimsuit. Here an analogy to the structure of shark skin was used bridging domains that are very far apart – biology and sports equipment. Applying bionics in product development offers very innovative and original solutions, but also difficulties in the transfer process.

The cases can be located along the efficiency / innovativeness continuum, coming up with three main clusters (see Figure 12.1). On the one end, we found three projects mainly focusing on reaching tight timeframes and cost limits and/ or a reduction of risks. On the other extreme of this continuum we identified two breakthrough projects that were mainly driven by finding a really new product solution. Finally, we found eight "Balanced" projects

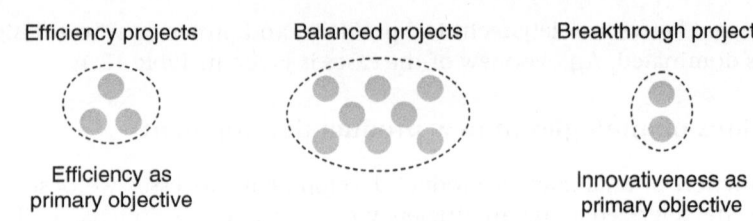

Figure 12.1 Project clusters according to primary function of using analogies

Table 12.2 Efficiency projects

No.	Target of project	Analogy	Characteristic of analogy based transfer
8	Mobile phone appliance for tunnels	Transfer of a processor used in a control board of an amplifier for a mobile phone mast	Transfer of an existing technological solution that was already used in a former project
13	Facility to rearrange allocation of train wagons	Transfer of a solution of consistent time measurement in shared systems	
5	Electronics for a medical device to create high voltage in very small dimensions	Transfer of a technological solution from light electronics	

which are located between pure efficiency and breakthrough. These are characterized by a compromise concerning efficiency and the goal of finding a highly innovative solution.

In the following sections we describe and analyze the various projects within the three clusters.

Efficiency projects

The three cases portrayed in Table 12.2 all belong to engineering companies that reported to use analogies mainly to meet time and cost constraints set by their clients. Hence, in these cases the search for analogies was mainly driven by the aim to increase efficiency. In order to succeed the engineers built upon their experiences from former projects and transferred existing technological solutions. For example, in case number 8, a mobile phone appliance for tunnels had to be developed. One of the participating engineers who was responsible for the hardware development made use of an already existing control board that he had developed in a former project as part of an amplifier for a mobile phone mast. He realized that he could use the complete processor for this new application. Due to this transfer of an existing technological solution the development time and costs could be reduced. In addition, existing contacts to the manufacturer of the processor could be used which had a positive effect on the procurement of the component. Altogether, the use of the analogy helped to reach the cost and time targets that had been defined by the client.

Relying on previous experiences can be highly efficient: In that case not only explicit knowledge, but also implicit knowledge can be transferred. Besides it helps in judging the relevance of the analogy if one has profound knowledge based on own experiences in the source domain.

Balanced projects

In all "Balanced projects" the reduction of development time and costs played an important role with regard to the use of analogies (see Table 12.3). At the same time all projects aimed to develop a really new solution to differentiate the innovation from existing products in the market. Thus, the common basis of the balanced projects is that a time and cost frame to develop a marketable solution is given by the client and within these constraints analogies are used to maximize the innovativeness of the solution. However, slight differences can be identified in-between the cases concerning their approach of making a transfer based on an analogy.

First, in the projects 7, 1, and 6 design elements or principles that had already been used in former projects were transferred. The transfers were not as direct as they were in the case of the "Efficiency projects," because only basic shapes and solution principles and not existing technological solutions were transferred. For example, in order to develop a new "baton" for the Commonwealth Games an analogy was detected to a digital antenna that had been developed in a former project. The antenna had the form of a stick with a swelling in the middle on which a display was mounted showing lights to visualize strength and activity of the received radio signals. In analogy to this antenna the baton was designed as a thin stick with a swelling in the middle containing a display that showed the pulse rate of the person holding the baton.

Next, in the cases 10, 9, 2, and 12 technical solutions were transferred, but these transfers were not solely based on experiences from former projects. Although the ideas that were followed originated from personal experiences of the team members, the teams were also willing to look into different areas to build up new competences if required. These projects targeted at developing a really new or premium solution that could be manufactured with reasonable effort and be introduced to the market in the foreseeable future. For example, in the development of a purifier for

Table 12.3 Balanced projects

No.	Target of project	Analogy	Characteristic of analogy based transfer
7	"Baton" for the Commonwealth Games	Transfer of form and technical ideas from an antenna for digital radio	Transfer of design elements or principles from former projects.
1	Design of a mixer for audio engineering	Transfer of solution principles from an ergonomic study of the interior of cars	
6	Design of a cockpit of a big passenger aircraft	Transfer of ergonomic principles from vehicle construction and chairs in general	
10	Purifier for dental tools (for example bur, polisher)	Transfer of technical solutions from high-pressure cleaners and premium car-doors	Transfer of technical solutions from other industries. Ideas are based on former projects, but not directly transferable without building up new competences.
9	High-quality baby stroller	Transfer of disc brakes from mountain biking and single wheel suspension from vehicle construction	
2	Design of a steering lawn mower	Transfer of solution principles from vehicle construction	
12	High-quality backrest of an office chair	Transfer of technical solutions from sports and medical technology	
11	Original promotion item with long lasting value	Transfer of material and movement of a piece of ship yard waste	Transfer of design elements based on direct contact with the environment of the designer

dental tools two analogies were used. First, to develop a cleaning mechanism, an analogy was detected to high-pressure cleaners that were already known in the market for cleaning tools. This approach originated from a participating designer who already had experience in the area of high-pressure cleaners. A second analogy concerned the door of the purifier. In analogy to high-class cars, an automatic closing mechanism was developed that draws the door shut if it is not properly closed or left ajar. The developed product was unique in design and handling. Most project goals were reached. Only the development costs were higher than targeted, because core competences for this apparatus had to be developed. Neither the knowledge broker nor the client possessed extensive knowledge in this area before the project.

Finally, a slightly different approach can be observed in case number 10. The client requested the designer to develop a promotional item with substantial benefit for the user. The client did not restrict the solution space. Just after receiving this job, the designer participated as a tourist in a harbor tour. There he encountered a hump of waste which stemmed from rivets punched by a shipyard. One of those small curved metal plates inspired him to develop a bottle opener as a skipjack transferring movement and material from the piece of metal waste. In this case inspiration was sought in the direct environment. On the one hand, this approach was driven by high efficiency: The product should be given away as a promotional item – therefore having limits concerning production costs. On the other hand high originality of the product was demanded. This case however stands apart from the other displayed cases, because the developed product was from a technical point of view very simple and his client provided only few restricting parameters for the project.

Breakthrough projects

There is a last cluster encompassing two cases where innovativeness was the predominant objective – almost totally neglecting project efficiency (see Table 12.4). In these two projects the intention was to develop a design study for imaginary product lines of the future not having the restriction to be directly marketable. However, these studies inspired the development of products being successfully brought onto the market.

In these cases the project teams chose a different approach for finding relevant analogies. The scope of analogies was broader than in the other two clusters – basic ideas are transferred from really different areas like, for example, nature or Stone Age habits. Analogical thinking does not lead to the direct transfer of a technical solution or material, but to get a profound understanding of solution strategy and transfer very basic design elements.

Table 12.4 Breakthrough projects

No.	Target of project	Analogy	Characteristic of analogy based transfer
4	Design study of a forklift truck	An egg as an archetype form to provide protection	Transfer of basic ideas and shapes from very distant areas. Use of very far analogies.
3	Design study of a mini cordless electric screwdriver	A hand ax (/ hand-wedge) from the Stone Age	

For example, in the case of the forklift truck study the designers started with the question of what is really important for the driver. Imagining themselves to be in the position of a forklift truck driver, the designers identified the importance of shelter. An analogy was detected to an egg as an archetype form to provide protection. Therefore, the cabin of the driver showed the basic shape of an egg. The study resulted in the production of a prototype that was not suitable for the market, because of high production costs and a too futuristic approach. However, this prototype served as inspiration for other forklift trucks that were successfully introduced into the market. Besides, the analogy helped to focus the project and to communicate the goals to be reached.

Looking back at the three different clusters, it can be summarized that most of the cases were attributed to the "balanced project" cluster. That is, the project teams followed mostly a mixed strategy to maximize the innovativeness of the solution concepts within a given time and cost frame. In addition, some projects clearly favored one function of analogical thinking: Either efficiency or innovativeness. It could be shown that the motivation of using analogies influenced the approach where to look for analogies and which kind of transfers to consider. In the projects dominated by efficiency considerations, the teams conducted the search for analogical problem solutions in a rather narrow space and primarily relied on personal experiences from former projects. These teams focused on finding an already existing solution that could be transferred without substantial development effort. In contrast, the more considerations of innovativeness dominated the use of analogies the further the teams went to look out for analogies. In those cases the teams did not only search for easy transferable solutions, but were also willing to build up new competences or merely to transfer some basic ideas or shapes.

Detecting relevant analogies

After the exploration of the different functions analogies can be used for in product development and the resulting search space we will now describe and analyze the process of detecting relevant analogies based on the results of the interviews. Thereby we focus on the importance of the human factor.

An ideal process model of using analogies derived from literature analysis is displayed in Figure 12.2 (Schild et al., 2004; Herstatt and Kalogerakis, 2005).

Premises to search for analogies

As the examined companies are service providers, basic goals and constraints of their product-development projects are set by their clients. Based on this information a project team has to be configured. Taking

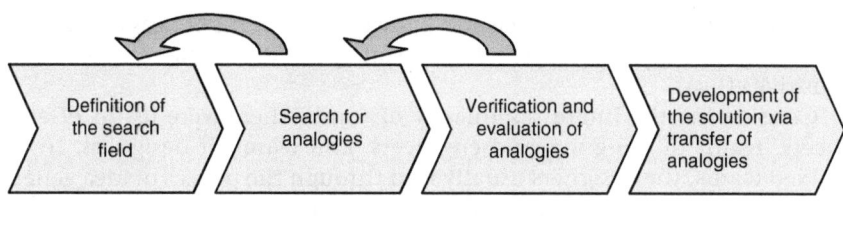

Definition of the search field	Search for analogies	Verification and evaluation of analogies	Development of the solution via transfer of analogies
• Abstraction of the problem • Review of constraints and general conditions • Integration of customer views	• Decision for a search strategy: search via people and/ or via databases • Use of methods to activate knowledge	• Was the analogue system correctly understood? • Evaluation of the analogy concerning its transferability	• Development in the company, in a cooperation or via outsourcing

Figure 12.2   Process model of using analogies in product development

into account that the search for analogies in the examined cases is mainly based on own experiences and personal contacts, characteristics of the team members seem to be decisive for the outcome of the processes. It can be valuable if former projects of the developers are considered while staffing the teams. In one of the engineering companies it is for example the job of the leader of the development department to choose the team members. He reported choosing the team members with the intention to increase the probability that experiences from former projects can be used for solving the current problem. If he realizes that there is no relevant knowledge in the company, he acquires external experts. The search of external experts and their contacting is again based on his personal network. However, from the interviews there is little indication that the consideration of increasing the success of analogical thinking strongly influences the team composition. Altogether, the formation of teams was predominated by general project management considerations. What kinds of skills are required? Who is the right person for contact with the customer? Who has the right determination for this project? Who is available?

For a knowledge broker it seems to be important that all employees participating in product- development projects possess individual characteristics fostering the use of analogies. First, according to the interviews, a person needs a certain amount of experience to rely upon. Therefore the building of diverse experiences through project work in a broad area of domains should be fostered. Only if people have knowledge from different areas are they able to make transfers based on analogies. Less experienced engineers or designers can continually broaden their horizon while working, if the teams are generally composed of people with different levels of experience. Second, communicative habits of the individuals are important. One has to be able to share his experience and knowledge with his colleagues. Finally, a certain curiosity, diverse hobbies, and an open view of the world can have positive effects.

Considering the interdisciplinarity of teams, there were teams of engineers, teams of designers and engineers, and teams of designers. In the mixed teams, the designers usually lead through the phases of idea generation and development of a basic concept. At the beginning engineers are mostly involved to guarantee the manufacturability of the solutions that the industrial designers create. Then, in later phases of the development of technically based products, engineers take over more responsibility. In the mixed teams, especially the designers are expected to look in diverse areas and get inspiration through analogies. However, the examples show that in the engineering-teams the use of analogies is also an essential part of product development. But engineers, probably due to their education, tend to look for simple transferable solutions close to the original area of the task.

If the project team is set it has to work on the project definition. The respondents indicated that a good problem definition at the beginning of the project is very important with respect to a successful outcome of the project. As indicated by the process model of using analogies, a project definition that includes an abstraction of the problem and considers given restrictions as well as the view of the customer views is needed to open a search space for analogies.

The need for an abstraction of the concrete problem could be confirmed by the results of the interviews. Abstracting the problem enables the developer to use knowledge from diverse domains. As the examined companies work as service providers they are not only confronted with the requirements of the future user of the product, but also the wishes of their clients. The demands of the future user are especially important in design projects. In order to consider ergonomic factors designers often perform human factor analysis at the beginning of a project. These ergonomic factors can be the basis for drawing analogies to former projects from different areas.

Altogether, a good communication – within the team as well as with the client – is an important premise in developing the project definition. Here, analogies can also help on another level. For example, designers sometimes use mood-boards to find a common understanding of the project with the help of analogies. Mood-boards are a form of visual stimulus: On large boards a collage is made with images that are usually cut out from magazines. It is used to help capturing the "values" of the product which will appeal to the target customer. According to its name, a mood-board should transport the mood of the product – "the sentiment, feeling or emotion which the product engenders when first seen" (see: www.betterproductdesign.net or www.aqr.org.uk/glossary/).

Search for analogies

The search for analogies is based on the problem definition and the identified general framework of the project. According to the interviews the search for analogies is not the outcome of an explicit and conscious decision, but emerges in the context of general creativity sessions held in the early phase of the development projects.

In general, as depicted in the process model, a search for analogies can be either based on knowledge stored in databases or on personal knowledge of the participating experts. None of the examined companies used databases to find analogies. The interviewees stated that the effort to initialize such a database and to fill it with new knowledge is too big. Furthermore, solutions in pattern matching to efficiently search in such databases were still not been sufficiently developed. One of the designers also explicitly mentioned a lack of time to search in databases or to execute thorough internet searches. Besides, he considered that a database-search for analogies was not a task that could be delegated to assistants, but would need to be performed by the participating designers and engineers themselves.

Altogether, database-search seems to be inefficient and is thus reducing the willingness to use analogies. Furthermore, only explicit knowledge can be transferred via databases. Implicit knowledge on the other hand has a subjective and intuitive character and is tied to persons. According to Swan et al. attempts to codify implicit knowledge of persons usually create knowledge that is useless, hard to verify, trivial as well as redundant (Swan et al., 1999). Therefore, access to knowledge while searching for analogies has to concentrate on the participating persons. Information technologies are only of secondary interest.

Experience and knowledge of the team members that is used to search for analogies can stem from diverse sources. First, former development projects are an important source. This type of direct solution transfer from former projects is especially relevant in the "efficiency projects" where a direct solution transfer from former projects occurs. A transfer from former projects also plays a crucial role in most "balanced projects." Other knowledge sources can be hobbies of the team members, their general education or an inspiration of the direct environment of the developers. In addition to applying personal knowledge, the team members can also make use of their personal networks. If none of the team members has personal experience in an apparently relevant area it might be that one of them has heard of a solution that could be relevant and knows where to find further information. The personal contacts of the team can lead to experts within or outside the company.

The question arises how the knowledge of the team is activated. According to the interviews some teams perform brainstorming or other creativity techniques (that is 6–3–5 method) under a given time frame. Although the formality of these creativity sessions differs, a discussion of diverse ideas within the team seems to always be an essential part of the search process. In addition, especially for designers, visual stimuli are important: for example looking into magazines, building mood-boards or trend-boards, studying other projects on the market or watching for inspiration in the direct environment.

The search for analogies is completed by an evaluation of the found analogies. It has to be checked if the analogue system was correctly understood and what kind of transfer can be made. In the examined cases the verification and evaluation of analogies is facilitated, because team members could refer to their own experiences in the analogue domain. This phase was mostly done in team discussions based on the project definition developed at the beginning of the project. At this point, it should also be considered if problems will arise through the transfer due to intellectual property rights or covenants with other clients. If this is the case a more abstract transfer might be a solution. However, in the given examples such conflicts did not appear.

Overall, the approach of the knowledge brokers to search for analogies follows no strict or formalized process. Several of the interviewees stated that their procedure of finding solutions needs to be very flexible, because each project possesses very individual characteristics. Furthermore, there is often insufficient time to follow a formal method or long and thorough search for analogies. This indicates that efficiency reasons often predominate. And the search for analogies usually is not identifiable as a distinct phase, but integrated into other approaches of finding a solution for the given task.

The most important factor seems to be experiences and characteristics of the participating persons. Therefore, human resource management has to be considered as an important factor. A diverse and broad knowledge of employees should be fostered if the company aims to use analogies in product development. Measures have to be taken to prevent the forming of specialists – employees of a knowledge broker have to be able to work to a certain degree as generalists.

Resumé and further aspects

All interviewees had positive experiences with the use of analogies in product development – designers as well as engineers. However, using analogies in their daily business is a more or less automatic process, because it is permanently part of the routine work of these people.

Analogies are used very pragmatically throughout the whole development process. First of all, they play an important part in the development of solutions. Here, we can differentiate between efficiency projects, balanced projects, and breakthrough projects. The function of analogies to increase project efficiency is an important result, because analogies are often just considered as a means to increase creativity. As shown before, the motivation to use analogies to create a new product directly leads to a narrower or broader search.

Second, analogies are also used for improving communication throughout the whole development process. They can already be used in the phase of project definition to develop a common understanding of the goals of the project. The project goals have to be communicated within the development team as well as to the client. Further, analogies can help not to lose the focus of the project and are valuable in discussions about possible solutions ("Shared picture or vision"). Finally, analogies can also be used to transport a message with the product to its future users. This is a main task of most design projects. Furthermore, as analogies are familiar to designers as well as engineers they may be especially useful in improving the communication between designers and engineers by decreasing the risk of misunderstandings that can arise due to their different backgrounds.

Finally, an important result of our research is that the search for analogies is mostly based on the knowledge of the participating persons. Therefore, human resource management has a crucial impact on the success of development projects. Design and engineering companies offering product-development services to diverse clients seem to automatically provide such a stimulating environment: Letting people build up experiences in diverse areas, supporting the personal communication throughout the company, but also to external experts, and using analogies as part of their daily work seem to have positive effects. These effects can also be seized by industrial companies if they transfer these mechanisms to their development departments.

Managerial implications

- Using analogies in product development is well perceived – by designers as well as engineers. However, using analogies in daily business routines today is a more or less intuitive, "automatic process", partly because it is part of the daily work of these people.
- In the cases we have observed, analogies are very pragmatically used throughout the whole development process. First of all, they play an important part in the development of solutions. Here, we can differentiate between efficiency projects, balanced projects, and breakthrough projects. The function of analogies to increase project efficiency is an important result, because analogies are often just considered as a means to increase creativity. As shown before, the motivation to use analogies to create a new product directly leads to a narrower or broader search.

- Using analogies is improving communication throughout the development process. They serve in the phase of project definition to develop a common understanding of the goals of the project. The project goals have to be communicated within the development team as well as to the client. Further, analogies help not to lose the focus of the project and are valuable in discussions concerning possible solutions ("Shared picture or vision").
- Analogies can also be used to transport a message with the product to its future users. This is a main task of most design projects. Furthermore, as analogies are familiar to designers as well as engineers they are especially useful in improving the communication between designers and engineers by decreasing the risk of misunderstandings that can arise due to their different backgrounds and "sticky-data" phenomenon.
- The search for analogies is mostly based on the knowledge of the participating persons. Therefore, human resource management has a crucial impact on the success of development projects: Letting people build up experiences in diverse areas, supporting the personal communication throughout the company, but also to external experts, and using analogies as part of their daily work seem to have positive effects. These effects can also be seized by industrial companies if they transfer these mechanisms to their development departments.

References

Bonnardel, N. and Marmèche, E. (2004). "Evocation Processes by Novice and Expert Designers: Towards Stimulating Analogical Thinking." *Creativity and Innovation Management* 13(3): 176–86.

Burt, R. S. (1992). *Structural Holes: The Social Structure of Competition.* Cambridge: Harvard University Press.

Dahl, D. W. and Moreau, P. (2002). "The Influence and Value of Analogical Thinking during New Product Ideation." *Journal of Marketing Research* 34(1): 47–60.

Geschka, H. (1992). "Creativity Techniques in Product Planning and Development: A View from West Germany." In S. J. Barnes Ed., *Source Book for Creative Problem-Solving.* Buffalo, New York: Creative Educational Foundation Press: 282–98.

Geschka, H., Moger, S., and Rickards, T. (Eds) (1994). *Creativity and Innovation – The Power of Synergy.* Darmstadt: Geschka & Partner Unternehmensberatung.

Gick, M. L. and Holyoak, K. J. (1980). "Analogical Problem Solving." *Cognitive Psychology* 12(3): 306–55.

Hargadon, A. (2002). "Brokering Knowledge: Linking Learning and Innovation." *Research in Organizational Behaviour* 24: 41–85.

Hargadon, A. (2003). *How Breakthroughs Happens: the Surprising Truth about How Companies Innovate.* Boston, MA: Harvard Business School Press.

Hargadon, A. and Sutton, R. I. (1997). "Technology Brokering and Innovation in a Product Development Firm." *Administrative Science Quarterly* 42(4): 716–49.

Herstatt, C. and Kalogerakis, K. (2005). "Haifischhaut als Vorbild für den Schwimmanzug." *io new management* 3: 26–31.

Herstatt, C. and Verworn, B. (2003). "Bedeutung und Charakteristika der frühen Phasen des Innovationsprozesses." In C. Herstatt and B. Verworn (Eds), *Management der frühen Innovationsphasen: Grundlagen – Methoden – Neue Ansätze*. Wiesbaden: Gabler: 3–15.

Holyoak, K. J. and Thagard, P. (1995). *Mental Leaps: Analogy in Creative Thought*. Cambridge: MIT Press.

Kim, J. and Wilemon, D. (2002). "Focusing the Fuzzy Front-End in New Product Development." *R & D Management* 32(4): 269–79.

Koen, P., Ajamian, G., Burkart, R., Clamen, A., Davidson, J., d'Amore, R., Elkins, C., Herald, K., Incorvia, M., Johnson, A., Karol, R., Seibert, R., Slavejkov, A., and Wagner, K. (2001). "Providing Clarity and a Common Language to the 'Fuzzy Front End'." *Research Technology Management* 44(2): 46–55.

Majchrzak, A., Cooper, L. P., and Neece, O. E. (2004). "Knowledge Reuse for Innovation." *Management Science* 50(2): 174–88.

Schild, K., Herstatt, C., and Lüthje, C. (2004). *How to Use Analogies for Breakthrough Innovations*. Working Paper: Technische Universität Hamburg-Harburg.

Swan, J., Newell, S., Scarbrough, H., and Hislop, D. (1999). "Knowledge Management and Innovation: Networks and Networking." *Journal of Knowledge Management* 3(4): 262.

Index